WINDSWEPT

WINDSWEPT

THE STORY OF WIND AND WEATHER

Mary de Villiers

WALKER & COMPANY
NEW YORK

First published in the United States of America in 2006 by
Walker Publishing Company, Inc.
Distributed to the trade by Holtzbrinck Publishers.

For information about permission to reproduce selections from
this book, write to Permissions, Walker & Company,
104 Fifth Avenue, New York, New York 10011.

All papers used by Walker & Company are natural, recyclable products made
from wood grown in well-managed forests. The manufacturing processes conform
to the environmental regulations of the country of origin.

Library of Congress Cataloging-in-Publication Data

De Villiers, Marq.
    Windswept : the story of wind and weather / Marq de Villiers.
    p.   cm.
    Includes bibliographical references and index.
    ISBN-13: 978-0-8027-1469-5
    ISBN-10: 0-8027-1469-2
    1. Winds.   2. Weather.   I. Title.

QC931.D48 2006
551.51'8—dc22   2005023115

Original book design for Sable Island by
Maura Fadden Rosenthal/mspace

Illustration on page 59 by William Gilkerson.
Illustrations on pages 4–5, 43, 76, 78, 83, 239 by Dereck Day.

Visit Walker & Company's Web site at www.walkerbooks.com

Typeset by Westchester Book Group
Printed in the United States of America by Quebecor World Fairfield

2   4   6   8   10   9   7   5   3   1

The wind goeth towards the south, and turneth about unto the north, it whirleth about continually, and the wind returneth again according to his circuits.   —Ecclesiastes 1:6

The wind being contrary, we betook ourselves to prayers again.
                    —Jesuit father, *Voyage to Siam,* 1685

Weather is personal.   —Forecaster's Credo

CONTENTS

CHAPTER EIGHT   229

The Technology of Wind

EPILOGUE   282

APPENDICES   287

CHAPTER ONE

Wind's Mystery and Meaning

*T*he story of Hurricane Ivan: It began, as
these things so often do, long ago and far, far away. Long ago, at least, in the
reckoning of weathermen, and far away at least as seen from the Caribbean
and the east coast of North America, where the storm's full fury would in
due time be unleashed. In the course of its tumultuous and destructive life,
the cyclone they came to call Ivan would exemplify all the perilous uncertainties and complex patterns of global climatology (and exaggerate my own
rather paranoid view of hard weather), but its beginning was hidden, even
secretive, and could only be seen in rueful hindsight.

In the spring of 2004, it rained in Darfur, the Sudanese hellhole wracked
by decades of civil war. Darfur is on the southeastern fringes of the endless
emptiness of the Sahara, and its soil, beaten down from too many cattle and
too many goats over too many years of drought, couldn't hold the water. It
pooled and then gathered in little muddy torrents that swept away the scattered huts of the countryside. A few days before, the refugees in their grim
camps had been dying of thirst—an ostrich egg of water having to do for a
family for a whole day—but were now forced to scramble to keep their pa-
thetic scraps of food and their meager possessions from washing away. They
were still starving, though now sodden and burdened with cholera and
dysentery in addition to their other miseries.

All along the Sahel, the southern fringes of the Sahara, the rains came.
Lake Chad, which had been shrinking for decades, stopped shrinking
briefly, and the remaining hippo channels winding through the papyrus and

water hyacinths filled up. The dusty plains north of Kano, the Nigerian trading city, looked lush for the first time in fifteen years. Outside fabled Timbuktu the ground took on a shiny green sheen, before the goats in their insatiable hunger nibbled the new plants down to a stubble, then trampled the residue into the mud. In Niger, Mali, even in ever-arid Mauritania, the rains fell for the first time in a decade. Not enough, really, to unparch the desert, but more than usual.

No one in the Sahel knew why it was raining, or, except for a few aid agencies, cared; they were just grateful the water was there. In the outside world hardly anybody paid much attention. There were a few exceptions—the paranoid actuaries for the giant insurance company Munich Re, for example, who are paid to worry, and a few analysts in hurricane centers across the Atlantic, who were wrestling with the complex causative cycles of violent weather—but more people should have been concerned than that, for they were about to get a brutal lesson in the interconnectedness of natural systems. Who would have thought that, say, a rural tavern in Pennsylvania would be threatened by a storm-born flood that was linked in complicated ways to the ending of a drought half a world away? But the green shoots peeping through the sand in the wadis near Timbuktu meant really bad news for the oblivious citizens of Florida and Alabama and the Gulf Coast of Mississippi and bad, though not quite so dire, news for the citizens of the eastern seaboard all the way up to Nova Scotia, where I live. They—we—would learn that in due time.

The search for an understanding of wind and the weather it brings has been a constant of human history, for wind is a changeling that can bring blessings but also hard times. Wind can be soft and beguiling, seductive; the caress of a gentle breeze stroking the skin is one of the great pleasures of the human adaptation to our natural world. But sometimes wind can be deadly, intensifying violently into a kind of personal malevolence. Like a short-tempered and belligerent god, the wind has a power that can appear arbitrary, excessive, overwhelming, devastating, uprooting trees, wrecking houses, sinking ships, battering people, scarring psyches.

At least, it can *seem* malevolent, and the malevolence can *seem* personal.

Out at sea at the southern tip of Africa near the Cape of Storms (which is what the Cape of Good Hope was called before the early colonizers' public relations flacks issued a "clarification"), the collision of two ocean currents sends massive pulses of disturbed air into the sky. The Benguela current, still chilled by the Antarctic frosts, and the Agulhas current, still humid with tropical warmth, intersect just southeast of Cape Town, and the storms they cause coil and twist, boiling up great black thunderheads, tearing the surface off the sea with a howling roar, and assaulting the land beyond. The gales race across the Cape Peninsula and blast out to sea again, across Table Bay to the open Atlantic beyond, where they finally lose their potency in the frigid waters off Namibia. On just such a day, on the spacious lawns of Sea Point on Table Bay, the winds seized a helpless child and knocked him down onto the grass, gratuitously, brutally, effortlessly. He struggled to his feet and yelled for help, but the gale snatched his breath away and blew it out to sea, stripping away the sound so no one could hear him, and the yell became a soundless scream. Then a gust punched him to the ground again and took hold and buffeted him toward the edge of the lawn by the shore, where a stone walkway skirted the breakwater and the rocks below, slippery with weeds and needle sharp with barnacles, pounded by breakers coming in from the sea. In the grip of the gale, the child skidded across the grass until he landed with a crack against the metal railings that were all that prevented him from being hurled into the ocean . . . It was there that my mother came, and fetched me away, and tried to still my terror with her beating heart.

It was a long time ago, but at that instant for that child and the man he became, wind indeed became personal, a *thing*, capricious

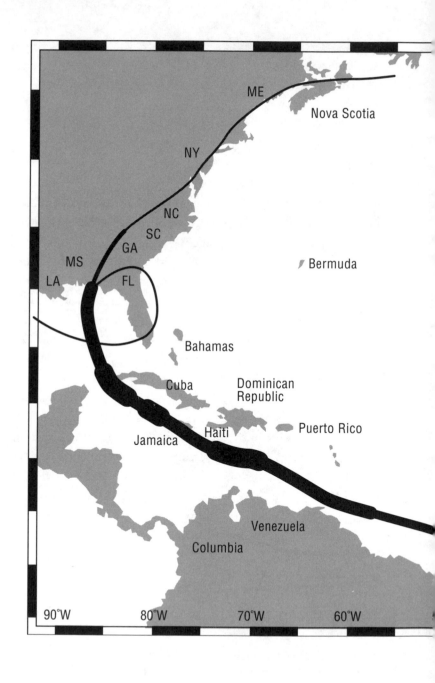

ME

Nova Scotia

NY

NC

SC

GA

MS

LA

FL

Bermuda

Bahamas

Cuba

Dominican
Republic

Haiti

Puerto Rico

Jamaica

Venezuela

Columbia

90°W   80°W   70°W   60°W

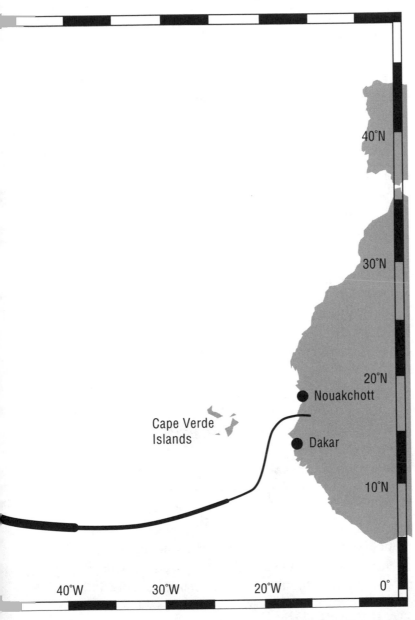

40°N

30°N

20°N
● Nouakchott

Cape Verde
Islands

● Dakar

10°N

40°W   30°W   20°W   0°

Path of Hurricane Ivan in the Atlantic Basin. The thickest portions of the track
are the points at which Ivan reached Category 5 status, the most powerful.

and malevolent, to be treated with the utmost caution and constant suspicion. I learned that unless you are very careful, wind can kill. And sometimes it will kill even when you take the most meticulous care. I learned that wind is perpetual, persistent, present even in its absence, in that sense, eternal. Wind has many disguises, is capable of many definitions. And possesses many talents, some of them cruel.

For years afterward, I was watchful, wary even of the gentlest seaside breezes. And now, perversely, I live on the Atlantic coast, in the teeth of the North Atlantic gales and in the ominous eye of the hurricane path. And of course, I'm still wary.

For some decades after my traumatic morning in Cape Town I lived mostly in cities, where I barely noticed the weather or the wind. Modern city people generally don't, I think. Weather is an occasional nuisance, but not something that affects life. Torrential rains come once or twice a year, occasional blizzards bring traffic to a crawl, gales can shake buildings and bring down trees, but really all you have to do is wait indoors for a while and it will all go away. True, heat waves and droughts are weather too, and if they persist and the water is rationed, they can seem alarming, but in the big cities of the developed world we have derived the reassuring notion that someone, from somewhere, will come along to fix it. Someone always does, if enough people grumble loudly enough. Even damage from the great northeast ice storm of 1998, which shut down power for millions of people in the dead of winter, was fixed after a while. People hunkered down and waited it out. Exceptions to this general obliviousness can be found, among them emergency workers, who must risk any weather to rescue the fools and punish the knaves who have ventured into it, and long-distance truckers, for whom weather is reduced to what bad weather perpetrates on a constantly shifting five-hundred-mile strip of asphalt, but as a generality it is true enough.

On farms, in the country, in small villages, and on the shore, the perspective is very different. People in those places have a personal, visceral connection to the weather and the wind. On my grandfather's farm on the arid plains of South Africa, to take one such place, the summer heat would boil up into the sky when the rains failed, the air would prickle and crackle and you could see the dust storms coming in from the west, a massive wall of violent color. You knew you had an hour or so to prepare, to close the windows and batten the shutters, to get the animals into the barn if possible, and then the sky would go black, shot with violet and brown, and the sand would blast the fruit from the trees and the flowers in their beds. And when it was gone, the heat lifted for a brief while and there were little drifts of sand by the doors and gates and everything was gritty to the touch, precious topsoil turned to dust and used by the weather to abrade whatever had stood in its path.

These years were the worst, the drought years. People hated the weather, then. And the wind as its personification.

On the east coast of North America the attitude is different, but the same. From our house early in the mornings I can hear, and sometimes see, the fishing boats setting out from the little harbor of West Berlin, a mile to the east. The fishermen have been up since four, and have checked the weather, but on most days the satellites and the forecasters with their sophisticated models will tell them what is already apparent. On days when the air is too clear, the black spruces across the bay too visible, branch and twig too obvious, the air calm but the sea feverish with oily swells coming in from the southwest, the fishermen know there is "weather" coming, driven by a system that started, perhaps, at Hatteras and is even at very long distances making the sea shiver. A hundred miles to the east, on the Sable Bank, the swells are mounding up four hundred miles in front of the storm, and they already pound the south beach of Sable Island, making that midsea dune quiver like a jelly with the weight of water assaulting the sand.

In these parts, the northeasters are the worst. When the satellites tell them a nor'easter is coming, the fishermen will go out in their little boats and haul their traps and gear, and then go home, drinking coffee in the kitchen until it is gone.

December 2004. Such a gale has just come, and departed.

Two days ago there was no sign of it. It was one of those wonderful days in which no real wind blows, only a gentle sea breeze. The sea in front of our house was glittering in the early-winter sunshine. The harbor seals were cruising just off the rocks, their heads silver in the brightness, and the ducks that had escaped the fall-season hunters puttered about in the ripples; I could see mergansers and eiders and mallards, and an occasional loon. We had an early-season snowstorm, but it was rapidly melting in the sun and a perverse but heady smell of spring was in the air, bruised spruce and bayberry. The breeze was benign, the sun smelled of winter's end here at the beginning of winter, the chickadees flittered about, even the porcupine that waddled across the yard didn't seem so damned obnoxious. I figured that when the wind is friendly, anything is possible, and "personal" no longer translates as "malevolent." There was no sign whatever of "weather."

But then I checked the forecast. At once, the computer screen lit up with a red flag—a wind and blizzard warning had just been issued. A low-pressure system was lying over the Carolinas and was forecast to track northeast and "intensify dramatically"—a fraught phrase from a forecaster, that—before reaching the Gulf of Maine. We would get lots of snow and gusts of wind well into the hurricane range. The usually matter-of-fact prose of the forecasters referred to "damaging winds" and "whiteout conditions." We were warned to monitor updates and to stay off the roads when the storm came.

I lifted my eyes from the screen and looked out the window. The trees were barely moving and the bland sunshine gave nothing away. One of our cats strolled uncaring down the driveway. Gulls cruised overhead. Nothing outside indicated the anxiety to come.

I got up to look at the barometer. The glass was steady. No change whatever. No sign that a roiling coil of supercharged air was churning into Pennsylvania, to be bent in its northerly course and steered our way.

But by the following dawn, sure enough, the fury was upon us. At the height of the gale I watched the spruces and firs from my office window. Their tips, often laden with cones, whipped violently about, and sometimes they broke, deadfall in the yard or the road, and would have to be cut up for firewood. The seas off our rocky beach were rearing up and hurling themselves on the shore with a thunderous roar. The swells were so high that they obscured the horizon before they broke—we figured later they must have been twenty-five or thirty feet high, tons of falling water that made the bedrock quiver under the onslaught. When the gale reached Force 10 and then, briefly, 11, the pitch of the wind rose from a moan to a shriek that picked at the nerves; then I shut down the computer—we would almost certainly lose power anyway, out here in the woods—and closed the shutters, and waited. My wife, who is braver, finds the wind exhilarating. I find it fearsome.

At least we knew it was coming.

Before the days of daily weather forecasting it was very different, and storms were much riskier. At sea, skippers kept a wary eye on the weather, but the storms sometimes came up with such appalling speed and ferocity that they were often taken unawares, with occasionally fatal consequences. They had to learn to trust their instincts, and to flee when they could or strike sail when it was too late. I remember a conversation with Fred Crouse, an old Lunenburg seaman who had served as Third Hand on one of the Grand Bankers as a lad, living through the twin hurricanes called The Gales of August of 1926 and 1927. He was off Sable Bank on the schooner *Partana* when the 1927 hurricane hit:

It was a fine afternoon, couldn't a been nicer. I was with Frank Meisner an' our bait was all [that is, finished]. So he said we'll try to git in Canso see if we kin git squid bait. There was no power then, just sails, and we hoist all the sails. Well there was just wind enough that she went along about three, four miles an hour. Before dark we was all turned in. He [the captain] come down an he said, "Fred, I never seen a sunset like this in my life. I can't believe it but we're goin' a have somethin' of this." He said, "Call the gang out an put the gear off the deck."

We all laughed at him and when we had the gear put in the hold I just said to him for a joke, I said, "What shall we do, batten the hatches?"

He said, "You better do."

"About the sails, how 'bout them?"

He said, "I t'ink you better haul down the mains'l an tie up the jib . . . put the storms'l on."

So we done that an' it was still fine weather. The crew all laughin' at about what in the name a the Lord he was about. Well nine o'clock that evening we wasn't sorry we done it! It come right the same as you emptied it out o' a bag. Oh, it blowed some bad.

The wind blew so hard that night that it ripped off the bait board and hurled it into the Atlantic—and this a two-inch-thick spruce plank nailed tight with sixteen five-inch spikes, torn loose by the wind. The topmasts had sheared off, the canvas was in tatters, their dories had vanished into the deeps, the helmsman was black from bruises. But the vessel survived.

It was fearsome bad when it blew, the old seaman said. If you had time, you could flee into the open ocean and ride it out with just a small kerchief of sail, and you'd likely be okay. Without warning, though, you could be caught in shoal waters, or on the windward of an island, with gear on the deck. That was trouble, big, big trouble. Many a man and many a vessel foundered and were lost.

We had warning of our storm because of those magician meteorologists in their weather bunkers, with their soothsaying devices, the Doppler radars and scatterometers and dropsondes and the rest, giving us the alerts we needed. Even so, despite the fact that we knew it was coming, and were braced for it, and understood what was causing it (warm air from the Gulf of Mexico colliding with cooler Arctic air, the whole system violently stirred and then steered by the jet stream), it was still hard not to feel somehow *targeted* . . . It really didn't seem fair. Why here? Why now? Why *us*? Fred Crouse in 1927 would have had no such warnings except for the canny instincts of his skipper, who had divined a pattern from subtle signals, and how much more arbitrary the storm must have seemed to him. And yet he was sailing in what was essentially the modern age, when radio receivers were already current and the national weather services of several countries were making educated guesses about what was happening in the wider world.

How much more terrifying still would storms have seemed in the prescientific ages, then? To Columbus, the first mariner we know of to survive a Caribbean hurricane. And to his predecessors, the Basques and the Vikings, the Phoenicians and the Greeks, the Chinese and the Arabs, who all ventured into the great emptiness of the ocean knowing nothing whatever of how storms were made. How much more malevolent when you had no idea how winds were generated, or that storms traveled in more or less predictable ways, or that there were natural systems that could be understood that caused storms to rear up . . . If all you knew was that great storms suddenly sprang out of nowhere, how very *angry* the world must have seemed, and how *arbitrary*. It's no wonder that wind and gods were conflated, and one took on the characteristics of the other—wind gods were generally angry and arbitrary gods, for winds were ferocious and ever changeable.

At least partly because of this conflation and confusion, wind plays a role in the creation myths of almost all human cultures, and in dozens the winds also govern commerce, procreation, communication, and more. For example, in the complicated cosmology of the Dogon people of Mali, in West Africa, there were four founding couples, thereby avoiding, in their view, the Christian sin of original incest; the women's mouths opened and the winds came out, and so did breath and therefore all subsequent life. The Mi'kmaq people of the American Atlantic coast have as their hero Glooscap, who once imprisoned the eagle whose wings create all the winds, and thereby made the world uninhabitable, albeit briefly. The San of the Kalahari believe the winds carry stories—not just legends, but also whole histories up to and including the very latest news and gossip. Old Japanese legends say the winds banished the fogs that shrouded the world in the Elder Days. The Bozo people, who live on the southern fringes of the Sahara, believe Wind wrestled with Water, and Water lost, which is the origin of the Great Desert. The Niger River is all the water that is left from the dawn of time, and the Bozo people became its stewards, taking up boatbuilding as a tribal preoccupation and sacred calling. In Hindu mythology, primordial winds swept bare the earth, preparing it for life. Even a casual pass through a biblical concordance yields a dozen or more references to wind—winds bring plagues to Egypt, winds collapse Job's house and kill his sons, God keeps winds in his heavenly warehouses, and a God-sent wind parts the Red Sea and lets the Jews escape from Egypt.

In many cultures winds are male and can impregnate unwary or unruly females. In African legends the fleetest antelopes are often wind-born as well as wind-borne. Hiawatha's mother conceived from the West Wind. In aboriginal legends winds originate in volcanoes, in caves in the mountains, from vents in the sea, from the breath of gods. Odysseus carried the four winds in a leather sack on his back and tied them to his mast (and his crew loosed the wrong ones, bringing ruin to his journey and giving Homer a great narrative line). The old Chinese god of winds also carried the winds in a bag slung over his back.

Sir James George Frazer's repository of old legends, *The Golden Bough*, recounts dozens of rites for mitigating, or at least controlling, the unruly winds. The Payuge Indians of South America used to light brands and run at the wind to frighten it; others pushed the winds into caves and rolled stones against the openings to seal them in. Frazer reports that Finnish wizards sold knotted ropes to mariners—the first knot for a breeze, the second for a stiff wind, the third for a gale, and the fourth for . . . well, you generally didn't want to untie the fourth knot, unless you were in port and your enemy was still at sea.[1]

As late as the nineteenth century Sir Walter Scott reported meeting an old woman in the Orkneys "who subsisted by selling winds. Each captain of a merchantman, between jest and earnest, gives the old woman six pence, and she boils her kettle to procure a favorable gale."[2] And stories are still extant in the memory of New Englanders, just yarns now, but colorful enough. Some were recounted by Richard M. Dorson in his collection of American tales called *Buying the Wind*. One told of the sea captain Paris Kaler, a notorious blasphemer. "Well, he got out one day and was becalmed, he was going to west'ard and there wasn't no wind. And he wanted some wind, so he throwed a quarter overboard. He wanted to buy a quarter's worth. So he said it commenced to blow, blowed till it blowed the sails off her, and he was three or four days off his course, he was three or four days getting back again. He said if he knew it was as cheap as that he wouldn't have bought half as much."[3]

In many cultures, too, gods came to represent the four cardinal directions of the wind. Most North American tribes believed the winds to be four gods. So did the Greeks, before they complicated things. Sir Edward Burnett Tylor in *Primitive Culture* suggested a commonplace explanation. "Man naturally divides his horizon into four quarters, before and behind, left and right, and thus comes to fancy the world a square, and to refer the winds to its four corners."[4]

The flat earth, in support of this view, was seldom a circle, almost always a square. It was obvious, then, why on Judgment Day, according to Revelation 7:1, four angels stand at the four corners, holding back the Four Winds "so that no wind would blow on the earth, or on the sea, or on any tree." In a nice admixture, "modern" flat-earthers incorporated into their design the devices of witchcraft, so that the earth had five corners, not four, being a pentagon. These five corners were to be found at the northernmost extent of Lake Mikhail in Tunguska, Siberia; on Brimstone Head, on Fogo Island, Newfoundland; on Easter Island in the Pacific; at Hokkaido in Japan; and somewhere near the south of Tasmania.[5] I've been to Fogo Island, where the local council has hospitably erected a rather vertiginous boardwalk to the top of Brimstone Head for those flat-earthists who want to see where one of the pillars is to be found.

In early Greek mythology the empire of the winds was shared between the four sons of Eos, the goddess of dawn, and Astraeus, the god of starry sky. They were called Boreas, the north wind; Zephyrus, the west wind; Eurus, the east wind; and Notus, the south wind.

Boreas, who dwelled in the mountains of Thrace, assumed the form of a stallion to mate with the mares of Erichthonius, and from this union were born twelve young mares so light of step that they ran across fields of standing corn without bruising an ear of grain and over the crests of the sea without wetting their feet. The Greeks liked Boreas because he had dispersed the fleet of the invader, Xerxes. Zephyrus in later years became a soft and beneficial wind at whose breath the spring flowers opened, but in the early myths he was savage and baleful and took pleasure in brewing storms and tossing the waves of the sea. From his tumultuous mating with the Harpy Podarge were born the two horses Xanthus and Balius, who drew the chariot of Achilles. When he calmed down in old age, he was given the gracious Chloris for a wife, by whom he had a son, Carpus, meaning fruit.

Eurus and Notus had little personality of their own, a reflection of the place of those winds in the Aegean.

According to Homer, the winds all lived in the Aeolian Islands, where they were kept under guard by Aeolus himself. The son of Poseidon, Aeolus was said to have invented the sailing ship, and Zeus appointed him guardian of the winds. It was Aeolus who gave Odysseus the wineskin containing the contrary winds that would hinder his voyage. Of course, Odysseus's greedy crewmen untied the bag to see what it contained, and let the deadly winds out.

Aeolus later became the Roman god of wind.[6]

It's a name that has persisted. Aeolian sound is the sound wind makes. We live all our lives in the Aeolian zone.

Virtually every part of the world has named winds—regular winds that the locals have personified over the centuries. Although no one knew where they came from or what caused them, winds were given names because, invisible and mysterious though they were, they were as real a presence as any mountain, river, or sea;[7] they were either benign—good for the crops, good for sailors—or, more likely, malevolent, the action of some malicious deity, sent to try the wretched people affected. Englishman Mike Ryding has tracked hundreds of them on his engaging Web site Whirling Winds of the World or, of course, WWW.[8]

Sometimes, indeed, these winds were kindly, life-sustaining, forgiving. One of the oldest extant references to wind in human literature is in the ancient Sanskrit poem Rigveda:

May the wind blow healing hither
Kind, refreshing to us in the heart,
May it extend our lives

Wind, you are to us a father
And a brother and our friend
So equip us for life

And if, Wind, there in your house
A store of immortality is laid,
Give some to us, that we may live[9]

And sometimes winds were like moody friends, occasionally rude but often charming. Sailors are unusually forgiving about their friends the winds, because at sea there "are no permanently ill winds. They are made for travel: one to take you away and, because no wind stays the same, another eventually to bring you back." Guy de Maupassant put a sailor's view very well in his essay *Sur l'eau* in 1888: "What a character the wind is . . . An all powerful ruler, sometimes terrible, sometime charitable . . . We know him better than our fathers and our mothers, this terrible, invisible, changeable, cunning, treacherous, ferocious person. We love him and we fear him, we know his tricks and his rages . . . He is the master of the sea, he who can be used, avoided or fled from, but who can never be tamed."[10]

If the winds were violent, they must be the work of some capricious god, quick to anger, a god who takes transgressions personally. As Joseph Conrad put it in *Typhoon*, violent winds can be like "the sudden smashing of a vial of wrath. It seems to explode all around the ship with an overpowering concussion and a rush of great waters, as if an immense dam had been blown up to windward. In an instant the men lost touch of each other. This is the disintegrating power of a great wind: it isolates one from one's kind. A furious gale attacks him like a personal enemy, tries to grasp his limbs, fastens upon his mind, seeks to rout his very spirit out of him." And again: "The gale's howls and shrieks seemed to take on . . . something of the human character, or human rage and pain."[11] "I was dreadfully frighted!" said Defoe's Robinson Crusoe. Homer has Odysseus reacting to Zeus's wrath: "The whole ship reeled from the blow of his bolt and was filled with the smell of sulfur. My men were flung overboard and round the black hull they floated like seagulls on the waves. There was no homecoming for them, the god saw to that."[12]

Shakespeare, as usual, has a trenchant word or two on the subject, and not just in *The Tempest*. Violent storms form the backdrop to three of his greatest tragedies: *Macbeth, Julius Caesar*, and *King Lear*. "Thunder disrupts the skies; shrieks of owls, laments and prophecies pierce the night air. The raging wind threatens to destroy the rooftops and, 'Some say, the earth, Was fevrous and did shake,' as Macbeth descends, dagger in hand, upon the sleeping Duncan."[13]

Certain local winds, like the *meltemi* that rules the Greek summers, are based on larger patterns. The *meltemi* is dependent on the monsoons that center on Pakistan in conjunction with the steady high pressure ridge over the Balkans. Winds like this are predictable in their patterns. "Aristotle, for example, said it arrived after the summer solstice on June 21. Once the *meltemi* blows a watch may be set by its rhythms. Stirring at ten in the morning, blowing hard around two, dying at six and calm by eight."[14]

Sirocco is the general name in the Mediterranean for the desert winds that blow from all across north Africa: "the *leveche* of southern Spain, the *chili* in Algeria, and the *ghibli* of Tunisia and Libya. In Egypt and the Levant the local name is *khamsin*, meaning 50 in Arabic, for the 50-day period the wind tends to blow. In Andalusia the south wind will parch a wheat field before it ripens, making the grain fall to the ground during harvest . . . but it is further north, after the *sirocco* has picked up Mediterranean sea moisture, that it becomes most noxious. The air takes on what Thomas Mann described in *Death in Venice* as a repulsive condition—an unbearably clammy, maddening state that saps energy and ruins the nerves."[15]

The Tuareg nomads of the Deep Desert call the *sirocco* the *harmattan*, the "hot breath of the desert." Sometimes the Tuareg and Tubu clans of the Ahaggar and Tibesti call the *harmattan* the *shahali* or *shai-halad*, the mother of storms, a phrase later picked up, to much ridicule, by Saddam Hussein. The wind's name is supposed to be derived from the Arabic word for *evil thing*; if it is, most people in the desert would agree it is well named. "When the wind blows the desert trembles," the Tuareg say, the dunes literally shiver and shift and horizons disappear. According to a chronicler in France's Foreign Legion in *Beau Geste*: "And across all the harmattan was blowing hard, that terrible wind that carries the Saharan dust a hundred miles to sea, not so much as a sand storm, but as a mist or fog of dust as fine as flour, filling the eyes, the lungs, the pores of the skin, the nose and throat, getting into the locks of rifles, the works of watches and cameras, defiling water, food and everything else, rendering life a burden and a curse." He didn't know the half of it: The *harmattan* brings out Raoul, the Drummer of Death, and his reach is greater by far than that of the Great Nothingness of the Sahara. Three hundred, four hundred, five hundred miles to the north, the hot breath of the desert commonly layers fine dust on Marseilles, on St. Tropez and Nice, and the swimming pools of the rich in the hills above the Côte d'Azur become filled with gritty milk. Dust from north Africa is also commonly found in northern Germany and England.

I've felt the effects of the *harmattan* myself. I once crossed over the border from the arid north of Cameroon to the arid south of Chad and its capital, N'Djamena, in such a blow. Before I even reached the city, half the Sahara seemed to be passing overhead, visibility was down to less than one hundred yards, and the air was thick with grit. The air had been stripped of what little moisture it contained; when the *harmattan* comes, humidity has been tracked to fall from 80 percent to 10 percent within hours. When the gale is in full cry, visibility is reduced to a few yards. Sand penetrates everything. Grit gets into the food, the water, on the sheets in the hotels. If you close the windows against the sand, as you must, the temperature can climb steadily, and reach 120 degrees, 125, and the air sears the lungs.

Many other winds in the desert have names: The dry northeast wind is called the *alizés*, and blows hot toward the equator. The dry, desiccating south winds that carry the glowering towers of dust are known variously as the *ghibli, chili, samun, jefhya,* and *irifi.* The *ghibli* is bad, and can seem never-ending. In the Fezzan of southwestern Libya, the camel masters say that if the *ghibli* blows forty days, God preserve us from the evil! The camel becomes pregnant without the intervention of the male. "Nothing can be more overpowering than the south wind, El-ghibli, or the east wind, El-shirghi, each of which is equally to be dreaded," British explorer George Lyon wrote in his journals in 1818. "In addition to the excessive heat and dryness, they are so impregnated with sand that the air is darkened by it, the sky appears of a dusky yellow and the sun is barely perceptible. The eyes become red, swelled and inflamed, the lips and skin parched and chapped, while severe pain in the chest is invariably felt in consequence of the quantities of sand unavoidably inhaled."[16] In the North African Campaign of World War II, several major battles were interrupted by the *khamsin.* Gales of 90 miles an hour and electrical disturbances so profound that compasses became useless forced the troops of both sides to hunker down, waiting for a lull.

In the Gobi Desert, the whistling winds were said to have been named for the sirens of the open sands, whose passion was to lure men to their deaths, but, of course, in reality it was the other way around: The sirens were invented to explain the winds.

On the northern side of the Mediterranean, especially around Greece, an even worse wind is the *gregale,* which blows from the northeast. "It was a gregale that drove Saint Paul from Crete to Malta in the Acts, and another that so undiplomatically interrupted a summit between Mikhail Gorbachev and George Bush."[17]

I haven't yet named the winds that swirl around our house, though they have personality enough. Perhaps I have yet to understand them properly. Perhaps they are not steady enough. Or rather, they are steady but too variable to take real shape. We had a local blacksmith make up a weathervane for us in the shape of a codfish, which we call Wanda, and yesterday Wanda started the morning in the southeast, swung to southwest by noon, and by midafternoon was pointing northwest. All in all, a fairly typical day—ocean breezes, land breezes, fronts drifting in from the west, disturbances coming up from what are still called "the Boston states" around here.

When the winds are southwest, they blow straight across the bay, heaping the water up in front of them. That's when the massive waves crash on the rocks with a sense-obliterating roar.

You can find this exhilarating, or you can find it intimidating. It depends partly on mood, but also partly on physiology, for winds can radically alter the body's heat-exchange devices, increase evaporation, and affect surface circulation of the blood; when they reach above about 12 miles an hour, almost everyone finds some discomfort. A number of scientific studies have suggested that human bodies are hardwired to the weather and that we are sensitive to shifts in temperature, humidity, cloud cover, wind speed, and barometric pressure.

No doubt it is possible to overcome this weather sensitivity and become inured to wind. If you're a sailor or a wind farmer—or a child with a kite—then doubtless psychology will trump physiology and you will find positives where others find only irritation.

This notion that winds affect health is hardly new. Hippocrates, the father of medicine, wrote a treatise he called *Airs, Waters and Places*, in which he observed that "whoever wishes to pursue properly the science of medicine must first investigate the seasons of the year and what occurs in them"—as quoted in Smith's *Southern Wind*. He urged his fellows to notice that warm southerly winds gave people "a humid and piteous constitution, and their bellies [are] subject to frequent disorders, owing to the phlegm running down from the head; the forms of the body for the most part are rather flabby. Women in such areas are prone to excessive menstruation, infants to convulsions, men to attacks of dysentery, diarrhea, and chronic fevers." For the unfortunates living in cities where west winds were common, the news was worse. People there were pale and enfeebled and subject to all the aforesaid diseases. North winds, for their part, induce "dullness of hearing, [and] if the wind prevails, coughs and infections . . . occur." Only about cities facing east could he be wholly positive.[18] In a *Discover* magazine article, Stephen Rosen's *Weathering* is quoted, listing a number of other eminent persons who believed that weather, and especially wind, was related to the body and controlled its humors—men of the world like Columbus, Charles Darwin, Ben Franklin, Johannes Kepler, Blaise Pascal, and Leonardo da Vinci.[19]

In more modern days, some scientists agree, sort of. Clinical psychologist John Westerfield has written that he'd seen a number of psychoses related to storm phobias. For a decade the German weather service has issued biometeorological bulletins like "cloudy with a chance of migraines and damp with a chance of insomnia." Michael Persinger at Laurentian University in Sudbury, Ontario, has conducted numerous biometric studies related to weather. The elderly, particularly, frequently cannot bring themselves to go outdoors because the wind makes them anxious.

The anxiety is worse when the winds are not just violent, but relentless too. In the dust bowl of the Dirty Thirties, a farmer's wife, Edna Jacques, penned a lament:

> The crop has failed again, the wind and sun
> Dried out the stubble first, then one by one
> The strips of summer fallow, seared with heat,
> Crunched, like old fallen leaves, our lovely wheat,
> The garden is a dreary blighted waste,
> The very air is gritty to my taste . . .

In that terrible decade, the rains had failed. Or, rather, the rains became normal, for over the centuries they had often failed and the landscape had adapted. But the farmers could not adapt. When prices collapsed and the drought persisted, the unanchored topsoil began to move. Blowing topsoil drifted across roads and railroad tracks, keeping the towns and cities bathed in dust and grit inside and out, causing yellow twilights at midday. The roads were impassable, with deep drifts of sand that built up until they covered the fences, choked out the few remaining shelterbelts and gardens, and reached the roofs of chicken houses. On May 12, 1934, the Associated Press reported huge clouds from the Great Plains dust bowl at ten thousand feet over the Atlantic, and amateur statisticians began calculating the amount of arable soil that had been removed in this storm alone, reaching more than three hundred million tons before the exercise became pointless. It was called the black blizzard, the name ironically foreshadowing the laments of Uzbek peasants a decade or two later, when their own landscape was blown away after the Aral Sea began to shrink, a disaster caused by overuse and careless engineering. The American black blizzard swept from the Rocky Mountain states to Washington and New York, and deep into the thoughts of Congress.

Many bitter tales from the Great Depression are obsessed with the winds that never ceased, banging the doors and shutters, rattling rickety barns, filling every crevice with drifting sand, a constant howling that picked at the nerves and seemed to cause violence in otherwise peaceable people. "My husband stood it for two months," a memoir by a Saskatchewan farm wife lamented, "watching our farm blow away in the winds, listening to that awful whistling, and then one day he hauled open the door and fought his way out onto the porch, yelling and screaming so hard it broke my heart." Her husband strode off into the gale and was not seen again; he disappeared into the wind and the wind-driven sand until he could walk no more, and then—or so it is assumed—the wind did him one final service, and covered over his body with the drifting soil of his own farm. He was neither the first, nor the last.

Jan DeBlieu's *Wind* mentions the classic West Texas melodrama *The Wind*, published anonymously by Dorothy Scarborough in 1925. Its dreamy eighteen-year-old heroine from Virginia wrestles unsuccessfully with the deprivations of mind, spirit, and body of life in endless drought and bitter poverty. It was the demon wind that was her undoing. Under its maddening influence her fragile spirit crashed. "With a laugh that strangled on a scream [she] sped to the door, flung it open and rushed out. She fled across the prairies like a leaf blowing in a gale, borne along by the force of the wind that was at last to have its way with her."[20]

Russian peasant leader Stenka Razin, about whom many legends swirl, claimed the maddening wind as his ally in his battles against the hated boyars, driving more than one enemy brigade to binges of suicides and mass self-mutilations.

During a *sirocco* in the Mediterranean, the oppressive heat is enough to drive up crime rates in Naples and Palermo. In Sicily, a *sirocco* that lasts more than three days is an excuse for a crime of passion. In America, in the eastern foothills of the Rockies, the incidence of rape is said to go up in a Chinook. It is said—though I haven't been able to prove it—that in Wyoming a law dating from the mid-nineteenth century allowed that wind-induced insanity was a sufficient defense against a murder charge.[21] One of Raymond Chandler's best-known Philip Marlowe passages in "Red Wind" deals with the Santa Ana winds that blow down the Cajon and San Gorgonia passes: "There was a hot desert wind blowing that night. It was one of those hot dry Santa Anas that come down through the mountain passes and curl your hair and make your nerves jump and your skin itch. On nights like that every booze party ends in a fight. Meek little wives feel the edge of the carving knife and study their husband's neck. Anything can happen."[22] Novelist Brian Moore, who lived in Malibu, used to describe the Santa Ana as stealing down over the hills, a

brazen thief in the night. Gabriel Garcia Márquez has written of
wind seeming a personal affront "aimed at us and us alone."[23]

Late in the summer of 2004 I talked to a couple of tornado chasers,
who spent a good part of the supercell season in Kansas, hoping
against hope to watch a tornado touch down close enough to feel. I
was convinced that this was part of the same suicidal pathology that
drove men and women to murder in the devil gales of the High
Plains, but they would have none of it. It was just an adrenalin rush,
they said, a high that comes from staring down one of nature's most
awesome and destructive forces. They were impatient with what
they saw as my senseless probing for deeper motive. For my part, I
wasn't convinced by their bravado. I had just finished a novel by Paul
Quarrington called *Galveston*, (*Storm Chasers* in the United States),
which was about a collection of misfits whose greatest desire was to
place themselves in the path of a Caribbean hurricane, and I thought
I understood them better than they did themselves. Modern folk are
supposed to be beyond this. In the old days, yes, weather was a grim
and capricious dictator, as the BBC's Felicity James pointed out,
quoting Penelope Lively's 1996 novel, *Heat Wave*. "But for the technologically literate, 20th century spectator, the weather is [supposed
to be] an aesthetic diversion."[24] I knew this was not so. My own ob-
session with watching the oversize blunderings of Atlantic hurricanes stemmed from the same source as the tornado chasers'. For
those of us who were no longer religious, it was awe. And not a little fear.

The mystery of wind's all-powerful pres-
ence, then, is deep-seated in the human psyche. It is one of the oldest mysteries of all, almost as old in our reckoning as the miracle of
the quickening of life and the awesome presence of the sun and the
moon. Humans in all cultures have been wrestling with winds and
their meaning from the beginning. Meteorology and astronomy are

the oldest of sciences, and in some ways the history of science is the continuing struggle to understand weather and its carriers, the winds. As the story of Hurricane Ivan shows, even now in the age of terabyte computers and chaos-driven algorhythms, explanations are still unfolding.

CHAPTER TWO

Wind's Great Theater

*I*van's story: Air isn't yet wind. The atmosphere is only the theater in which wind presents itself, and air is only the stuff of which wind is made, and at first this thing that was taking place in the southern Sahara, this not-yet-even-pre-Ivan, was merely a matter of moving air masses, a ponderous vertical circulation of the unseen and unknown, in a place still far, far away.

Far away from us, at least. It was late summer, and we had visitors from Paris at our house. One balmy afternoon we went down to Beach Meadows Beach, as our local white sand crescent is called, and strolled along the tide line, looking for shells and signs of clams. Our friend Philippe called it "votre petite paradis" (your little piece of paradise), and so it was, because the beach was empty and warm underfoot, and the sea glittered in the sun. There was no surf to speak of, only small ripples in the water. There was a gentle breeze aloft, in which the gulls played.

But on that same day, half a world away in the Sahara Desert not far from Darfur and in a landscape alien and hostile even by Saharan standards, something altogether different was happening.

This was in the massif called Tibesti—"Tu," or rock, to its inhabitants, the Tubu. It is no small thing, covering about 300 miles northeast to southwest, and about 175 across, smaller than the more famous Saharan Ahaggar Mountains but rivaling those in the ferocity of its demeanor. The massif lies mostly in Chad, but it stretches north into Libya and even west into Niger. It is really less a mountain range than it is a rupture in the earth, formed *when lava streams forced their way through the planetary skin, leaving a stratum of crystalline rock covered with sandstone. This sandstone was then forced violently upward again as two massive tectonic plates collided, deep within the mantle; it now rises abruptly from the surrounding plains, reaching to 11,204 feet at the summit of Emi Koussi, the Sahara's highest point. Some of the volcanoes are still active and emit a sulfurous smoke from the roiling magma far below. The whole place is deeply eroded, cracked with ravines and awful wells in the rock, unplumbed and of unknown depth. In places the lava resembles the waves of a petrified sea, albeit a sea with swells reaching five hundred feet or more. Once, in the Sahara's distant past, great rivers roared through these ravines, leaving scars that are still clearly visible. Three of these ancient river courses provided the secretive Tubu their way into the massif, and there they built a few scattered towns, among them the remote Bardai.*

On that summer day, so benign on America's eastern seaboard, the hot breath of the desert flung itself against Emi Koussi and scaled rapidly into the upper troposphere. The superheated air rushing up the slopes (in what is called anabatic, or up-mountain, flow) at some point met the cooler, denser air flowing downward (katabatic flow), creating pulses of turbulence, rapid mixing, and the formation of massive thunderheads and ominous, towering black clouds. The sulfurous volcanic air was riven by lightning, and thunder rolled across the ravines, echoing off the boulders that were scattered like a giant's abandoned toy box across the landscape. The system spun off tornadoes, their evil spirals twisting among the ravines. For the next few days the thunder cells drifted slowly westward, driven by the prevailing easterlies of the season. At high altitudes there were violent showers and localized mountain floods; on the superheated plains west of the salt mining center of Bilma the rains fell, but the air was so hot it flash-evaporated the water before it could reach the ground, and the Tuareg nomads could see the cooling water falling without ever feeling it. A day or so later the system passed south of Agadez in Niger, where it was reinforced by yet more turbulent air generated by the lower but nevertheless forbidding black peaks of the Aïr massif of the central Sahara. Then the set of thunder cells crossed the desert north of Timbuktu,

and the small weather station there recorded its passage in a handwritten
logbook. It was one of many that passed by that month. The records say
there were violent downdrafts and gusts of 60 miles an hour.

Air, in a gesture of atmospheric alchemy, had by now transformed itself
into wind. It was the last week of August 2004.

I

The first time I really thought about air, the stuff that makes the
wind, was on a beach on the Indian Ocean side of South Africa,
near the tidy little town of George. Our family was lying on the
white sand drying out after an early morning swim, and my cousin
Colin, who had the day before caught a clawless rock lobster by the
simple expedient of diving down through six feet or so of clear
water and picking it off the sand, asked a typically unexpected
question.

"If the crayfish can't see the water," he said, "and they look up,
do they think the fish are flying?"

There was a gentle onshore breeze that morning, I'm sure—
there was always a gentle onshore breeze in the morning—but I
didn't notice it. I remember looking up at the puffy cumulus clouds
scudding overhead, but I paid them no mind either. I found Colin's
question curiously vertiginous. It was similar to the question asked
in many a school geography class ("If you couldn't see the water,
would you not think ships were flying?") but in a way much more
unsettling. Did that mean we, like the lobsters, were merely bottom-
feeders? Living at the bottom of a towering pile of . . . *something* . . .
a great weight of something unseen and unfelt, roiling around
above our heads, and above that the clouds, and above that, what?
Where was the "surface" of the air? Was there some atmospheric
equivalent to the ground?

These were a small boy's ponderings and soon faded. But behind

Colin's simple question, I know now, was something much more profound. That we are indeed living at the bottom of a towering, restless sea of air goes some way to explaining the nature and persistence (but also the fragility) of life, the biosphere that inhabits the troposphere, here in the unsettlingly narrow layer between the bleakness of outer space and the unrelenting pressure of the boiling rock deep beneath the mantle.

We experience little of this directly, except as wind and weather. But where does wind begin? With the pressure differentials between highs and lows? With the solar energy that causes those differentials? With the nuclear fires in the sun that cause the solar energy that causes the pressure differentials? Or merely with that substance that was so invisible—and then later so mysterious—to our ancestors, the thing called "air"?

Wind, after all, is just air in motion.

Was there ever a time when there was no air? Very likely not. It may have been poisonous to us—but there was *something* . . . What exactly it was is still subject to speculation, which is the scientific word for guesswork. However, it is probably true that Earth has had three successive atmospheres with varying degrees of stability.

The most common assumption is that Earth itself is around 4.6 to 5 billion years old, formed by gravity from cosmic junk, clouds of ionized particles around the sun, and debris left over from the somewhere-sometime explosion called the big bang. This cosmic tip-heap coalesced to form a protoplanet, which grew by the gravitational attraction of even more junk, what the cosmologists call particulates. This was the so-called Hadean Eon: a sort-of Earth existed, but there were no continents, and no oceans—and most definitely no life. Just boiling clouds of gas.

At first, this not-quite planet was almost certainly too hot to retain much of the primitive atmosphere it was born with. Based on what we know of free gases in the universe, this first atmosphere would probably have consisted of helium and hydrogen. Until recently, the scientific consensus was that both these gases would have boiled off into space, to be replaced as the planet cooled with the products of volcanic outgassing—water vapor, carbon dioxide, carbon monoxide, sulfur dioxide, nitrogen, ammonia, methane, hydrogen sulfite, and chlorine. A new study, however, has found evidence that hydrogen persisted in the atmosphere, escaping into space much more slowly than earlier thought. It's possible this secondary atmosphere would have contained almost 30 percent hydrogen. There would have been no oxygen at all.[1]

The massive amounts of water vapor expelled from the condensing Earth would have formed a dense cloud layer, which then precipitated out as pure water.

Which raises the questions: Where did the water vapor in the atmosphere and in the condensing Earth come from in the first place? What was it in the volcanism that caused our first weather to produce H_2O? Very likely, it was already present in all that cosmic junk—comets are sometimes little more than frozen lakes of water—and it came from space, an alien and infinitely curious little molecule. For water *is* curious, much more curious than it might at first appear, and is actually little understood. Why is it, for instance, that water is the only substance whose solid form is less dense than its liquid one, a phenomenon that has profound implications for aquatic life? "As a liquid, water has special thermal features that minimize temperature fluctuations. First among these features is its high specific heat—that is, a relatively large amount of heat is required to raise the temperature of water. This capacity to absorb heat has several important consequences for the biosphere."[2] Water and wind intersect in important ways to regulate our planet.

Over a period of thousands of years, the rain accumulated as rivers and lakes and ocean basins. This left carbon dioxide, CO_2, as the main component of air. About half a billion years after its birth, give or take an eon or two, things had settled down enough for this carbon dioxide to react with water and other compounds to form rocks and minerals. The oldest known rocks are in Greenland, and have just celebrated their 3.9 billionth birthday. Earth was still aflame with volcanoes and bombarded by asteroids, meteorites, and whatever else was floating in the interstellar void and intersecting with our nascent planet, but those oldest rocks show signs of having been deposited in an environment already containing water.

At about the same time, methane began to increase. By our planet's billionth birthday, the atmosphere would have been 80 percent methane and related carbon compounds and less than 10 percent nitrogen. Nitrogen, which is not very chemically active, continued slowly to accumulate.

Nothing about this was unique. In the solar system the only body other than ours with an atmosphere is Saturn's largest moon, Titan. NASA's probes have found that Titan's "air" consists largely of nitrogen, hydrogen sulfide, and methane, with trace amounts of other gases—in effect, Titan's air is what we on Earth would call thick smog. Titans might thrive on it, but to us it would smell vile. Breathing it would quickly kill us. You can still get the sense of it by leaning (carefully) into a volcanic vent, if you do so between eruptions.

II

It can plausibly be argued that the most extraordinary event in Earth's long life, at least from the point of view of species now living, occurred when the planet was an adolescent, about three and a half billion years ago. This was the sudden, and still quite mysterious, appearance of photosynthetic bacteria, phytoplankton and algae.

These were not the planet's first life forms; yet more primitive sulfur-eating bacteria already existed, and indeed still do exist in miasmic sulfur vents in obscure places around the globe—bacteria that, for lack of a better theory, are thought to have emerged in a pool of soupy, chemical- and nutrient-rich water, although a small but influential scientific subset believes life too might have come to us from space, ready-formed, cosmic nuggets among the infinite dross. "Runoff collected in a small volume is the most likely means of achieving the necessary concentration of ingredients," Gustaf Arrhenius, a geochemist at Scripps Institution of Oceanography, told *National Geographic* in March 1998. The presence of hydrogen in substantial concentrations would help to explain how these concentrated compounds attracted other molecules that then acted as catalysts in subsequent reactions. Hydrogen is a volatile and active gas, and in its presence sugar phosphates, possible precursors to organic life, would have been produced.

In any case, the new photosynthetic organisms began the laborious job of converting carbon dioxide to oxygen and of splitting water vapor into its component parts, and free oxygen began to trickle, molecule by molecule, into the air. As Richard Fortey put it in his book *The Earth: An Intimate History*, "Three billion years of photosynthesis, much of it achieved by very simple organisms like blue-green bacteria . . . the minute rod- or thread-like remains of the bacteria that changed the world have been found among the earliest fossils. They formed sticky mats, which left their record in finely layered, crimped or cushion-like fossils called stromatolites, looking like piles of petrified flaky pastry." If you were alive at this period, Fortey wrote, "[you would see] a hot spring bubbling and hissing at your feet, surrounded by brilliant orange and purple livid stains. They feel slick to the touch. The colors are painted by bacterial life that forms a slimy film; the lively hues are pigments that shield the tiny cells from harmful rays in the harsh sunlight. From here, you can see the tacky green mats in the sea that are slowly transforming the earth's atmosphere."[3] At the same time, the days were getting longer and longer as the Precambrian progressed and the planet's rotation slowed, allowing ever more photosynthesis to take place.

Then, a mere billion years ago, the first eukaryotic organisms, the earliest ancestors of all of us, appeared. These were more "organized" cells than their predecessors, with DNA coding segregated into a nucleus instead of being jumbled up with everything else, and they were far more efficient at converting carbon dioxide to oxygen. The percentage of oxygen consequently began to climb, going up from mere traces about 600 million years ago to its present level of about 20 percent.

These processes, acting sequentially and then simultaneously, produced the delicate balance preserved in modern air, a combination of permanent gases (nitrogen at 78.084 percent of all air, and oxygen at 20.947 percent) and gases considered to be variable (gases that have changing concentrations over a finite period of time). Variable gases are essentially water vapor and aerosols (tiny liquid droplets that include ice crystals, smoke, sea salt crystals, dust, and volcanic emissions, suspended in the air), plus carbon dioxide, methane, and ozone, all of which are critical to maintaining planetary temperatures and therefore life (see Appendix 1).

This combination is not at all magical, except for the one simple fact that our lives are entirely dependent on it, and even a small deviation in its proportions would kill us and profoundly alter the planet's climate. But as Richard Fortey pointed out, nothing in evolutionary history made the photosynthetic bacterium that resulted in us "a dead cert" to emerge at all—the evolutionary winner could just as easily have been a bacterium that depends on sulfur, in which case we would never have evolved. Something else would have, but not us: No wonder evolution makes creationists nervous. It is not so much that we are descended from primeval slime that is disturbing, but that our whole existence is such a fluke.

Before the coming of life, then, Earth was a bleak place, a rocky globe with shallow seas and a thin band of gases—largely carbon dioxide, carbon monoxide, molecular nitrogen, hydrogen sulfide, and water vapor. It was a hostile and barren planet. This strictly inorganic state of Earth is called the geosphere, which is itself made up of the lithosphere (the rock and soil), the hydrosphere (the water), and the atmosphere (the air). Energy from the sun relentlessly bombarded the surface of the primitive Earth, and in time—millions of years—chemical and physical actions somehow produced the first evidence of life: formless, jellylike blobs that could collect energy from the environment and produce more of their own kind. Heritability was the real key to evolution, the ability to reproduce and therefore evolve. Thus was created the biosphere, the zone of life, an energy-diverting, entropy-fighting, rapidly evolving but perilously thin skin on Earth's surface that uses the matter of the earth to make living substance. It happened in water, and it happened in air. Air was the necessary predisposing factor, for all life depends on it.

III

So much we understand now; we can parse air to a fare-thee-well and measure particulate matter to a part per billion. But for millennia (from the beginning of recorded thought) and until very recently, air was the most mysterious of, well, *substances*, if that's what it can be called. In fact, there was considerable doubt that air existed at all; and when philosophers imagined it into being, there were no instruments able to measure it or prove its existence.

The first conceptual breakthrough came when the Greek philosophers grasped that air was *something* rather than *nothing*. It was, after all, far from self-evident. You couldn't see air, taste it, or measure it. True, there were clues—odors were a puzzle, and so was wind. Fog also—clouds were too far out of reach to worry about, but you could walk through fog, its appearance and its evident moisture obvious, though its nature was still obscure. Before that, air was just . . . nothing. We were like the crayfish of the Indian Ocean in my cousin Colin's question—you couldn't think about it because you didn't know there was anything there to think about.

The man who understood that things might be a little more complicated was Anaximander, a philosopher of the sixth century B.C. The sky and the earth and everything on it, he taught, were conjured into being when the great primordial sea was evaporated by celestial fires; everything left over is the consequence of basic oppositional forces—light and dark, dryness and wetness, heat and cold. He said nothing about air specifically, but his pupil Anaximenes took the discussion one significant step further. He suggested that air was itself an element, and the most important one at that: "Air is the first principle of things, for from this all things arise and into this all are resolved again." Its properties were intermediate between fire and water, and air was therefore the basic matter from which these two were formed. Some centuries later Plutarch, who was a great admirer of Anaximenes, quoted him as saying, "all things are [thus] generated by a certain condensation or rarefaction of it."[4]

The notion was obviously current in the schools of philosophy.

Anaximander

Among the earliest surviving texts, dating from about the same period as Anaximander, is a theory contained in a document called the Derveni papyrus, discovered in 1962 half burned on top of a coffin; it had been part of the dead man's funeral pyre. Derveni's tale was an allegory about Orpheus, but in the text the author's worldview was clearly stated. The world had two unequal basic principles, fire and air, and air is, at the same time, divine and also called "mind." Cold air has the property of checking hot fire. Originally, fire was dispersed through the universe, created havoc, and prevented the formation of order, or *kosmos*. To set creation in motion, mind-air acted to concentrate fire in the stars and the sun: This made the world as we know it.

This was, even by ancient standards, pretty hit and miss, and there it rested until it was taken up by Aristotle (384–322 B.C.), the great synthesizer of Greek thought. The meteorological and chemical ideas codified by Aristotle remained "true" for more than 1,500 years. They can be summarized as two overweening concepts: the four elements of matter and the atomistic view of matter. These overlapped and reinforced each other, but were also to some degree in conflict—Aristotle himself agreed with the former but not the latter, and the debate he set off was the source of more than a thousand years of sometimes acrimonious philosophical and alchemical wrangling.

The first complete description of the four-elements theory dates back to a philosopher called Empedocles (490–430 B.C.), but his notes were restated more elaborately and concisely by Aristotle. All matter, in this view, was made up of four basic, irreducible elements: earth, air, fire, and water. In turn, these four basic pillars of the universe are derived from the four "properties," hotness and its opposite, coldness; and dryness and its opposite, wetness.

Fire and water are obvious opposites, according to Aristotle, and so are earth and air. They have nothing in common and share no properties. Each element existed *somewhere* in an ideal, or pure form, not found on Earth. Real or earthly things were impure mixtures of the ideal elements. Smoke, for example, was a mixture of the air and earth with some of the element of fire added. The elements could be changed into one another by removal of one property and addition of another—an idea later seized on by the medieval alchemists, the precursors of modern chemistry. Another point that seemed obvious was that the four elements had a natural tendency to separate in space; fire moved upward, away from the earth, and the earth moved inward, away from the air. Air and water, for lack of any other conceptual framework, were described as intermediate. Matter, in whatever form, could be subdivided indefinitely in theory, although Aristotle acknowledged that this would not always be practical or easy to do.

It was, in fact, an early theory of everything, still the holy grail of science. It explained almost all the actions of nature. A fire, for example, was on earth merely impure ideal fire. When a pot was placed over a fire, the bottom of the pot became black; this happened because the real fire was a mixture of ideal fire and ideal earth, and therefore when the fire entered the pot to give it more of the property of hotness, some or all of the earth mixed with it was left behind as a soot. When sea water was heated, it absorbed the hotness of fire and moved away from water, becoming air; the impurity in real water, earth, then was left behind on the bottom of the pot as dry salt. And air, when cooled, would condense droplets of water, as when cold metal was placed in contact with the air above a kettle of boiling water or in moist air. This occurred because the property of coldness, taken from the metal or earth, moved the air toward wetness and therefore partially toward water.

Hmmm . . . it worked too. Mostly. But as a refinement, Aristotle, an honest skeptic, was obliged to add another element to the mix proposed by Empedocles. He called it *quintaessentia*, and the very name is a giveaway that he wasn't completely satisfied that the Big Four were a complete set. Quintaessentia was an eternal and unchangeable element otherwise known as ether, or space; it was the framework in which the other four existed, and it haunted physics until after 1850, when meticulous measurements finally made its retention impossible.

Aristotle also bought into the notion proposed by his master, Plato, that each of the four elements existed in a particular geometric form and the properties of the element were therefore related to that form. So fire particles were tetrahedrons, four-faced figures whose sharp points gave speed and burning sensations like arrows striking the flesh. Earth particles had the shapes of cubes, which accounted for their solidity; water particles had the smoother shape of twenty-sided icosahedrons, while those of air had the shape of an octahedron, a figure with eight sides. Ether, being the highest of the elements, had the most complex geometry, that of a pentagonal dodecahedron, a solid figure with twelve equal pentagonal faces. This was all wrong, of course, but not so wrong that its central notion, that each of the elements was made up of particles having a single definite shape, didn't resonate strongly with the modern theories of chemistry being developed in the seventeenth and eighteenth centuries.

The other prevailing Greek theory of the universe, and therefore of air, was the curiously modern-sounding theory of atoms. The first sighting of this interesting notion was in the writings of Leucippus, of whom otherwise nothing is known, and his student Democritus, somewhere around 400 B.C. They theorized, contrary to the four-elements school, that matter was not capable of infinite subdivision but contained ultimate and extremely small particles they called "atomus." These tiny particles, like matter, are eternal. Differences between substances are therefore due to the atoms of which they are composed, which are of different shapes and arrangements. Differences in the properties of substances, moreover, are not due to the atoms themselves but to the way in which they are arranged. The second point was that these atoms were in constant motion. The third was that they were separated by voids, or vacuums, in which they moved.

Aristotle disagreed vehemently with this whole idea of atoms. There was no such thing as a vacuum, he believed. Air has mass, yes—he showed that a container filled with air could not also be filled with water—but it couldn't possibly have weight; he flattened an airtight bag and weighed it, then filled it with air and weighed it again, and found no difference. And how could atoms be in constant motion? The whole theory, in his opinion, brought philosophy into disrepute.

Philo of Byzantium, in the third century B.C. performed a more sophisticated version of Aristotle's mass experiment, and was thus the first to really prove that air had substance. He attached a tube to a glass globe, then inserted the open end of the tube into a dish of water. When he placed the globe in shadow, the water rose within the tube. When he exposed the globe to sunlight, the level fell. "The same effect," he wrote, "is produced if one heats the globe with fire."

He had—though he didn't know it—stumbled on the true cause of wind.[5] He was the first meteorologist.

IV

And there the science of air rested, for another two thousand years or so. Inquiry into the physical world went on, but at a desultory pace, and philosophers turned their attention to other matters, such as the transmutation of base metals into gold, and to increasingly odd notions of cosmology; alchemy and astrology dominated the physical sciences until the Middle Ages. The first comprehensive theory after Aristotle's that sought to explain air and combustion wasn't formulated until the late seventeenth century. This phlogiston theory was entirely plausible in the light of current knowledge, though it, too, was entirely wrong. Its ascendancy lasted about one hundred years, and when it was finally put to rest, the way was opened for the birth of a modern, measurement-based, technologically oriented, practical discipline that we now know as atmospheric science, of which meteorology is an important part.

The theory was German in origin, and the two scientists most identified with it were Johann Joachim Becher and Georg Ernst Stahl, who first used the word *phlogiston* in 1700. Following the ancient Greek codification practice, the theory sought to reduce everything to three essences that make up all matter: sulfa, or *terra pinguis*, the essence of inflammability; mercury, or *terra mercurialis*, the essence of fluidity; and salt, or *terra lapida*. Terra pinguis was later renamed *phlogiston*. Under the pressure of the church, living matter was excluded from this taxonomy, because of course it contained the potential for a soul that differed in composition from nonliving matter, and since it was divinely inspired, was necessarily outside earthly classification. This was Stahl's vitalism theory, outlined in his *The True Theory of Medicine*.

The governing idea of phlogiston was that all metals were made of a calx, or residue, combined with phlogiston, the fiery principle, which was liberated during combustion, leaving only the calx. Air, according to this theory, was merely a receptacle for phlogiston. Combustibles, or calcinable substances, were not elements at all, but merely compounds containing phlogiston. For example, rusting iron was believed to be losing its phlogiston and thereby returning to its elemental state.

Similarly, flames extinguish because the air becomes saturated with phlogiston. Charcoal leaves little residue after burning because it is nearly pure phlogiston. Mice die in confined spaces because the air becomes saturated with phlogiston.

The theory had many advantages. It explained, for example, how air at first supports combustion and then after a while does not. It also addressed some of the obvious drawbacks of Aristotle's theory, particularly his hazy notions of chemical change. But phlogiston nevertheless came under increasing attack. At first, the results of hundreds of practical experiments in dozens of laboratories around Europe were shoehorned into the theory, but too many just didn't seem to fit. Why, for example, did some metals, such as magnesium, actually gain mass when burned? Phlogistonists finessed the discrepancy by assigning phlogiston a negative mass, or by asserting that air entered the metal to fill the vacuum after phlogiston left, but this satisfied hardly anyone.

Even Joseph Priestley, the towering and cantankerous intellect who discovered the existence of oxygen, was a devout believer in the phlogiston theory (he simply called oxygen "dephlogisticated air") and late in his life issued a ringing denunciation of the antiphlogistons, whose activities he actually compared to Robespierre and the Terror, a gross libel and a calumny. (This was from his new home in America, whence he had been driven by a lynch mob sent by King Charles, angry at his presumed anticlerical activities; Priestley was never one to suffer fools lightly—it was perhaps why he became such a good friend to Thomas Paine.)

It was left to the ever-practical Frenchman Antoine-Laurent Lavoisier, to demolish the phlogiston theory entirely. He was the first to understand the significance of Priestley's writing on oxygen, and he disproved phlogiston by showing that oxygen is required for combustion, as well as for rusting and respiration. Lavoisier, who began his career in the Royal Gunpowder Administration in Paris, is best known for his synthesis of chemical knowledge in his *Traité élémentaire de chimie*, in which for the first time the modern notion of elements was laid out systematically.

In the revolutionary spirit of the time, Lavoisier made a symbolic break with the theory by burning all the books on the subject he could lay his hands on.

And in the other spirit of the time, though a political moderate, he died on the guillotine during the Terror, in 1794.[6]

Lavoisier was influenced by Robert Boyle (1627–1691), who had made precise measurements in studying the relationship between volume and pressure of gases. In the *Sceptical Chymist* Boyle questioned Aristotle's view of the four elements and proposed matter was composed of tiny particles, therefore becoming the first modern to synthesize the two Greek notions of the universe. He was followed

by Joseph Black (1728–1799), who discovered carbon dioxide in
1750 and showed that it was produced in combustion, human breath,
and fermentation; and by Henry Cavendish (1731–1810), who found
that common air was made up of nitrogen and oxygen in a 4:1 ratio.
Carbon dioxide might still be called fixed air, hydrogen inflammable
air, and nitrogen dead air, but oxygen was no longer called dephlogisticated air, and Cavendish's mix was about right.

So now we knew what air was.

V

But we didn't yet know what the atmosphere was. That had to wait
for modern science, for the atmosphere is much more complex than
was thought even a hundred years ago. Air, with its patented mix of
gases and in its various degrees of thinness and density, is one thing,
but the atmosphere is also made up of bands of radiation, clouds of
ionized gases, zone layers, and swirling magnetic fields, all of which
impinge on our weather and our winds, and thus on us, down here
at our little intersection of atmosphere and lithosphere where living
things make their home.

The first complication is that the atmosphere has layers, according to the density of the air. In the conventional taxonomy, weather
occurs and we live in the troposphere, a layer of air ranging from
the earth's surface, where it is densest, to somewhere between 5
miles at the poles and a little more than 10 miles at the equator.
Within the troposphere, the air gets colder the farther away from
the earth it is, which is why airplanes flying at around 20,000 feet
usually push through air whose temperature is well below zero, even
at equatorial latitudes. (The range is generally thought of as "earth-
normal" to −60° Celsius). Despite the commonsensical observation
that winds are largely horizontal, which is how we perceive them,
the prevailing movement of air within the troposphere is actually
vertical, in what are now called Hadley and Ferrel cells.

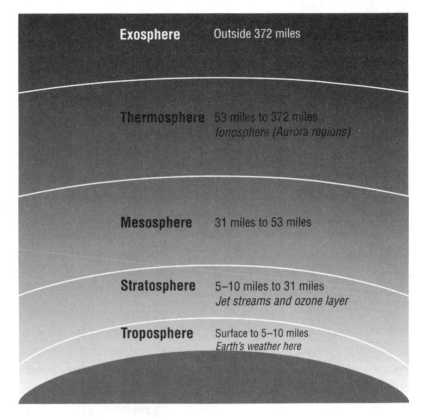

The atmosphere, showing the approximate ranges of the major layers, the aurora regions (in the thermosphere, at left), and the orbiting altitude of the space shuttle.

Above the troposphere is a thin layer of stable temperatures generally called the tropopause, to distinguish it from the next layer up, the stratosphere, which itself reaches to around 31 miles from the earth's surface. Temperatures within the stratosphere actually rise with altitude, reaching the freezing mark of water (0° Celsius) at the highest elevations. Stratospheric winds are almost all stable, persistent, and horizontal. This is where the now notorious ozone layer is to be found.

Above the stratosphere is—big surprise, this—the stratopause, and above that is the mesosphere, reaching up to about 53 miles. The air in the mesosphere is too thin to circulate much at all, and winds hardly exist there. Not too thin, though, to burn up most of the meteors that enter our atmosphere. Temperatures in the mesosphere decline, from about 0° Celsius to somewhere around −100° Celsius.

Then, of course, there's the mesopause.

Beyond that is the thermosphere, which stretches out as far as around 370 miles and represents the outer limits of earth's thermal reach. Temperatures in the thermosphere range from around −80° Celsius to as low as −1,000° Celsius, though the number of air molecules at that altitude is so small, and the consequent heat transfer so meager, that it would not feel at all cold to the human skin (provided that skin's owner could somehow deal with the absence of oxygen and the vacuumlike pressure).

Just to complete the set, however, atmospheric scientists generally include one more layer, which they call the exosphere. As its name implies, the exosphere is simply that part of space beyond any influence from Earth; the exosphere, therefore, is a near vacuum, containing not much at all, and yet includes pretty much all the remaining universe—the human species is a parochial one, measuring influence by its own small corner of its own small galaxy. The satellites that beam TV signals into your living room circulate on the lower fringes of the exosphere.

The air gets thinner the higher you go. Unsurprisingly, then, atmospheric pressure—and the density of air's life-sustaining gases—diminishes rapidly with altitude. At 9,000 feet the air pressure is already only three quarters of that at sea level, and almost everyone—except hardy Andes dwellers, Sherpas of the Himalayas, and a few Ethiopian and Kenyan marathoners—feels the effects. The reduced pressure causes the brain to swell slightly, resulting in headaches and nausea. At double that height, 18,000 feet, the pressure has dropped to half earth-normal; no permanent settlements exist at these altitudes because the human lungs just can't cope, and expel too much CO_2, fatally disrupting the body's balance. At Everest's peak, just over 29,000 feet, pressure is only 30 percent, about 300 millibars instead of the earth-normal of 1,000; at those altitudes, the human body begins an inevitable, and final, breakdown.[7] It's possible—at least for a few extrahardy mountaineers—to survive for brief periods at Everest's summit without an oxygen tank, but survival at that altitude is measured in minutes, not hours.

It's interesting to contemplate what a perilously thin layer our sustaining atmosphere really is. Earth's diameter is only a little more than 7,000 miles. The troposphere, in which our weather (and the greenhouse effect) happens, is at most 10 miles deep. Put another way, if the earth was a ball 4 feet or so in diameter, the troposphere would be a fraction of an inch thick, about the thickness of the lead in a common pencil.

$\mathbf{B}$ut the theater of the wind is more complicated than just stacking decreasingly dense layers one on the next.

In the popular imagination, and in the scientific consensus before the space age, "outside" seemed empty, dark, and frigid, a void through which to view distant planets and stars; the very word *space* seemed imprecise, lacking in definition, an afterthought, coined only in opposition to nonspace, or Earth. In fact, our planet's hinterland is not empty at all, but filled with magnetic fields, electric fields, matter, energy, ionized gases, and radiation, generally invisible to the eye or the telescope but easily apparent to instrumentation. Some of these are worth noting because they affect our climate and our wind patterns and therefore the weather.

The most curious of these is the ionosphere, a region starting about 50 miles above the surface, where atmospheric gases are so thin that free electrons can exist for brief periods before being captured, or ionized, by free-ranging positive ions, which in turn are produced by solar X-rays and ultraviolet radiation. The ionosphere is organized, more or less, into a series of broad bands, or levels, which for convenience sake but with scant regard for felicity are unimaginatively called D, E, F, and Topside. The D region, the lowest, and Topside, which is pretty obviously the highest, have little impact on human life; both are regions of weak ionization. It is the middle two layers, E and F, that impact us most. E is sometimes known as the Kennelly-Heaviside layer (or just the Heaviside layer), after the American engineer Arthur Kennelly and the British physicist Oliver Heaviside. It is a layer of strongly ionized gas between 54 and 90 miles thick, and because of its frequency (eight hertz), it has the useful effect of reflecting medium-wave radio transmissions, thereby allowing Midwestern country-and-western stations to be heard over the horizon in, say, Maine, to the great benefit of listeners there pining for hurtin' music; I remember in the old Soviet Union how teenagers would use the effect to listen to the Beatles singing "Back in the USSR" on the BBC, to the great irritation of their politically correct elders. Its usefulness varies, though, by time of day, season, and sunspot activity, so radio stations are wise not to make extravagant promises. The F region, above 90 miles, does the same thing.

But apart from its impact on radio broadcasting, why pay it any attention? Because the ionosphere is a flexible, dynamic, and rather fragile system buffeted about by electromagnetic emissions, by variations in the earth's magnetic field, and by the acoustic motion of the atmosphere itself, which means it is acutely sensitive to atmospheric changes. To monitor the ionosphere, then, is a good way of keeping a watch on atmospheric events. But there's more: Many scientists believe that, as NASA puts it on its Web site, "there is persuasive evidence of an ionospheric precursor to large earthquakes," and that it could even be used as a predictor. This is partly because acoustic waves are generated both before and after earthquakes, but also because it is thought that part of the run-up to an earthquake is the generation of electromagnetic emissions, which have been detected in the ionosphere up to six days prior to a large quake. In other words, if we learn how, we might be able to use the ionosphere as an early-warning system, dramatically increasing the time people living in earthquake zones have to react. A few days' warning could have made a massive difference, say, in the death toll of the Asian tsunami the day after Christmas 2004.

Above the ionosphere is the so-called magnetosphere. It is simple to visualize but complicated in its effects (which include, curiously, huge numbers of conspiracy-based Web pages that accuse the scientific community and the U.S. government in particular of massive cover-ups and fraud regarding its existence).

The magnetosphere is the field of influence of Earth's magnetic force—Earth is the equivalent of a normal, if rather over-size, bar magnet, mostly because of its nickel-iron core. Its most obvious use to us, of course, is in the magnetic compass; sailors have navigated by compass since it was independently invented sometime both in China and in Europe in the twelfth century, when savants in both places noted that a piece of lodestone (magnetic ore) floating on a stick in a jar of water pointed to the polestar. The magnetosphere ranges thousands of miles out into space—many orbiting space probes have as one of their scientific missions to help map this field, a task still incomplete. This simple and orderly system is complicated by the massive amounts of solar plasma making up the solar wind, which is itself strongly magnetized and pulses off the sun as a stream of high-energy electrons and protons at the speedy rate of about 240 miles a second, or about a million miles an hour (London to Paris in about three quarters of a second), bleeding off 10 million tons of hydrogen and helium from our sun every year—a trivial amount, considering its mass. This solar wind reaches Earth in about two to three days, before passing on through the solar system. It hits us in a spiral shape, mostly because of the sun's own rotation. This spiral has been given the homely moniker "the sprinkler" because it rather resembles the spray pattern of an ordinary lawn sprinkler. The sun's magnetic influence also pushes Earth's own magnetic field into a cylindrical shape, facing away from the sun—the solar wind compresses the sunward side of the magnetosphere downward close to Earth, to within six to ten times the radius of the earth, and conversely drags out the night-side magnetosphere to perhaps one thousand times Earth's radius. This end of the spiral is called the magnetotail.

The sun's magnetic energy doesn't penetrate all the way to the planet's surface, because one magnetic field cannot easily penetrate another. Earth's own magnetism deflects that of the sun, thereby helping to keep us alive. If our planet's core was made of, say, aluminum, we'd all be dead.

Within the magnetosphere are two doughnut-shaped radiation belts, sometimes known as the Van Allen belts after NASA scientist James A. Van Allen, whose Geiger counter on the probe *Explorer I* discovered them in the 1950s.

The inner belt, which circles Earth above the equator, is populated, in the scientific jargon, by high-energy electrons, mostly caused by cosmic rays, that readily penetrate spacecraft and that can, on prolonged exposure, damage both instruments and the people who use them. This belt is not quite a neat circle; the offset between Earth's true north and magnetic poles causes it to reach downward to about 150 miles above the Atlantic near Brazil, causing the South Atlantic Anomaly, a kind of Bermuda Triangle of near space. Low-orbiting satellites frequently pass through it, and are frequently damaged and sometimes fatally zapped as a consequence.

The outer belt is less hazardous but also less stable, and subject to electronic and magnetic storms. It tends to wax and wane with the sun's eleven-year sunspot cycle.

These Van Allen belts, it seems obvious to the conspiracy buffs who populate the Web's wilder shores, make it perfectly clear that the penetration of space has never happened, that man has never been to the moon, and that the Russian space station is really just a set built in the Nevada desert by Russian Communists and their American stooges. If gamma rays lurk in the belts and can kill astronauts, then obviously you would need tons of lead shielding to survive them, and so just as obviously no one has ever been through them alive—QED.

The facts, as they usually are in these cases, are rather less dramatic. It's true that both manned and unmanned spacecraft tend to stay out of the Van Allen belts if they can, but it is equally true that traveling at speed through the belts would yield a radiation exposure of about 1 rem (at 25 rem you start to show symptoms; at 100 you are dead). The principal hazard in the belt is not cancer-inducing gamma or X-rays, which readily penetrate most substances, as Superman could have told you, but high-energy electrons and protons, which are not difficult to shield against. In addition, the belts span only about 40 degrees of Earth's latitude, 20 each side of the equator, and so if the path of a spacecraft were inclined 30 degrees to Earth's equator, it would bypass all but the edges of the belts.

So, as with the ionosphere, why should we care? Astronauts need to care, but do we? How do the radiation belts affect the air, and therefore the winds, and therefore the climate, if at all?

We know that magnetic storms have caused current surges in power lines, causing blackouts. They also cause static interference that interrupts radio and television signals and cause dead zones for cell phones, to the annoyance of teenagers everywhere. More seriously, they cause air and marine navigation instruments to behave erratically. They can also change—and damage—the ozone layer that absorbs harmful ultraviolet radiation from the sun.

But it goes beyond this. It is becoming clearer that the sun's activity affects short-term weather patterns and perhaps long-term climate trends. Current thinking is that the solar wind has only a minor long-term effect on climate, although no one really knows, and a great deal of earnest activity is underway to uncover the truth. But with the wind and the weather, the evidence is rather different. As NASA's Dr. James Green puts it, "Changes in the magnetosphere seem to be transmitted to the lower atmosphere, where they may influence the circulation of air masses. If we can discover the physical links between these two regions of our environment that trigger weather and climate changes, we can better predict and prepare for our weather."[8]

Thus the theater of the air, the stage setting for wind. Wind itself is a product of solar radiation, but it is shaped and affected by planetary rotation, pressure differentials, and by air in all its multiple movements, patterns and paths. And air in turn is affected by ionization, radiation, magnetism, and the cosmic wind. No wonder the ancients found air ineffable and wind mysterious.

CHAPTER THREE

The Search for Understanding

*I*van's story: *All the thunder cells caused by the great furnace of the Saharan summer are tracked by the one-man weather office in Timbuktu and by the slightly more sophisticated operation in Niamey, Niger's capital. Timbuktu's sole meteorologist, Bandiougou Diallo, observes the weather the old-fashioned way—by going onto the roof in a thunderstorm, eyeballing its extent, and hoisting aloft a handheld wind velocity meter. He is there mostly to warn the pilots of Air Mali's venerable Fokker aircraft on their thrice-weekly runs to the city from Bamako, Mali's capital, if it is safe to proceed or more prudent to turn back. But his handwritten notes, forwarded later to his bosses in Bamako, are assembled into broader databanks and used by others to track storm patterns, and thus become part of the global struggle to understand weather systems. Not all the storms he watches cause him or his distant correspondents much concern. Some dissipate locally. Others lose their energy after a day or so. Others persist. A few coalesce into violent weather systems big enough to alert American meteorologists who are monitoring satellite images across the Atlantic.*

All thunder cells, though, travel, for that is the nature of the air that has become wind that has transformed itself into a storm.

If you were on the high dunes north of Timbuktu, the wind would have come from the northwest but you could have watched the storm approaching from the east. Saharan storms are easy to see: Sulfurous clouds of yellow sand swirl up into the angry blacks of the clouds. Probably no one was there to see, though. The nomads know better than to be caught on a high dune in a windstorm; they would have taken shelter, such as it was, in the lee of a smaller dune or in a wadi (though mindful of flash floods).

The cell that began at Emi Koussi passed by Timbuktu in the afternoon of August 27. It was one of those whose coherence persisted, and it drifted slowly south of west, passing unrecorded over the ancient capital of the Malian empire, the ruins now known as Koumbi Saleh, and was picked up again by weathermen on the 29th, somewhere between the Mauritanian capital, the arid desert town of Nouakchott, and the sprawling and violent slums of Dakar, in Senegal. Airports in both cities kept a wary eye on its passage.

By the end of the day, it had reached the coast. Ahead in its path: the Cape Verde Islands.

This was a complicated place for any weather system to find itself. Behind is the immense furnace of the desert. To the north, aridity. To the south, the sodden hills of Senegal, and beyond them the rain forests beginning at the Gambia River. To the west, the still cold but rapidly warming southbound Canary current, heading for the tropics. The prevailing winds—northeasterlies. Sometimes these factors simply kill thunder cells. At other times, whose conditions look apparently similar, they are energized as they are caught up in the southbound winds.

More than a hundred such systems drift out of the Sahara into the ocean each year.

When they reach tropical waters, they either stall, or they don't. Again, if the conditions are right—the water temperature just right, the upper atmosphere still so no shear is found at the high altitudes to cut off their tops—the low-pressure systems they have become begin slowly to spin as the Coriolis force takes hold.

The Emi Koussi cell was one of these. In the ocean south of the Cape Verde Islands the Canary current had already reached 28.8° Celsius. The high-altitude winds were steady at low velocities. It was a formidable combination. The effect was like taking the lid off a pot of steaming water—moist air began to ascend rapidly, moisture and energy were dumped at high altitudes, *surface winds rushed into the vacuum thus created and in turn forced the winds into a tighter circular motion.*

By this time the National Hurricane Center in Miami had taken notice. The cell was now formally a tropical depression, and as such it was assigned a number. It was the ninth of the season. Tropical Depression Nine, with sustained winds still under the tropical storm threshold of 39 miles an hour, was located 555 miles southwest of the Cape Verde Islands. It was September 2.

I

I was once pursued by a windstorm across the arid Great Karoo of South Africa. I'd been spending a few days with a friend at his aunt's house in the little town of De Aar—I mostly recall being stuffed by that hospitable woman with all the traditional delicacies of the Afrikaner heartland, grilled baby lamb, roast springbok, *pannekoek* and *moskonfyt, babootie,* and the rest—and we had headed back to Cape Town on Willie's knockabout motorcycle. The first we knew about the storm, the first inkling that something was amiss, was when our own dust overtook us; a following wind had gotten up, and it was strengthening fast. I looked over my shoulder and the horizon was already a luridly dirty purple lit up with sheets and jagged spears of lightning. For a moment I thought I saw a yellow-tinged funnel shape reaching toward the ground, and if it were true, that would mean really bad news. "Get on with it, Willie!" I said, and he opened the throttle and we made a run for it.

The machine juddered over the ruts. I could feel the storm behind me, *chasing* . . . This was stupid, I knew even then. It was just a storm, the basso rumbling that followed us just thunder, the flashing just lightning, the hail just ice, the wind just wind, and the massive clouds of debris—thorn bushes, small birds, dust and grit, tumbleweeds—was aimed at nothing at all. I knew that. I was no longer a child, to be battered by every passing gale. But phobias seldom pay court to rationality, and trump volition every time.

Why was the damn thing following me?

For three hours the storm nipped at our heels. The bike wasn't very good and the gravel road was very bad; we were making no better than 40 miles an hour, and the storm was keeping pace, traveling at furious speed across the veld. Even Willie, a structural engineer by training and temperament, felt its malice, or so I fancied. Finally, after hours of bone-jarring flight, the arrow-straight road veered left, and when we could see we were no longer in the direct path, we stopped the bike and watched the monster pass.

That the storm was moving was evident. We could see it coming after us, we could see it pass by, we could see it vanish over the horizon. So why, I wondered in later years, did no one figure out, in the centuries past, that weather traveled? The Tuareg in the desert could watch storms pass, mariners at sea could surely see how they moved. And yet, in two thousand years of musings about winds, by some of the brightest intellects in history, the notion that storms were self-contained "systems" that moved from one place to another was never mooted. It wasn't until the nineteenth century, when data collection was fully developed, that it was finally understood.

Why wasn't it evident to the ancients?

I suppose the answer is to be found in context. The natural philosophers of the past just didn't—yet—have the conceptual framework to understand how wind worked. This is the familiar story of the scientific method, of course. It is the way scientific understanding advances. First come the framing theories, almost always wrong, then the painstaking accumulation of data, then corrected theories, then more data, and only afterward the brilliant insight. After which, the pattern repeats itself . . .

The history of thinking about wind paralleled the thinking about air itself. It was similar, but not the same.

The first of the framing theories, the first quasi-scientific definition of wind, was that of Anaximander, the same Anaximander whose considered view of air (that every thing, earth and the heavens above, came into existence when the primordial sea was dried by celestial fires, source of said fires unspecified) was recounted in the previous chapter. All nature, in this view, was the product of a few simple properties—hotness and coldness, wetness and dryness, lightness and darkness. So far, so familiar. But then he made a big conceptual breakthrough: Wind, he suggested, was a current formed when mists were burned off by the sun. Later, he simplified the notion: Wind, he said, was "a flowing of air."[1] Two hundred or so years later Empedocles, the inventor of the four-elements theory of matter, conducted the first experiment to demonstrate what wind really was. He used a simple flow tank to show that air and wind were really the same thing.[2]

Aristotle, writing three hundred years later, agreed with much of what Anaximander had to say about air—indeed, he codified and improved the earlier man's musings—but on wind he couldn't go along. He rejected the whole idea, just as he had the notion of atoms. He thought it self-evidently false, and that the people who adhered to the theory were bringing the whole idea of philosophical thought once more into disrepute—obviously, scholarly dignity was a sensitive subject in the Athens academies. If winds were indeed a "flowing of air," he wrote, this must mean that all winds were one wind because all air was one air. To him, this was self-evidently false. "This is just like saying all rivers are one and the same river, and an ordinary man's view is better than a [mediocre] philosopher's view like this one," he declared, with rather lumbering irony.[3] Still, he wasn't altogether off the mark when he wrote that the sun pushes the winds and checks their speeds; in a way, that is indeed what happens. You can say with some truth that wind is, as Aristotle wrote, an exhalation arising from the earth.

While this philosophical wrangling was going on in the Athenian schools, practical Greeks were making the first attempts to blend legend and fact. In this they were the global forerunners, as they were in much else besides. Homer and his contemporaries had only identified the four cardinal winds, but this was too crude a measure for decent navigation, and the increasingly skilled sailors of later generations began to parse the directions ever more finely into increasingly useful and accurate segments, generally using the sun's movement as a guide, since magnetic north was still unknown. This new precision was graphically depicted in the Tower of the Winds, an eight-sided building erected in the market place of Athens, now the Plaka district, by Andronicus of Cyrrhus, somewhere between 100 and 50 B.C. The tower still stands, in decent though not pristine shape, and the eight winged gods representing the eight important winds can still be seen as a marble frieze, the figures in relief. The tower is only about forty feet tall. Sundials protrude from all sides (it was also called the Horologium, or timepiece) and a water-powered clock was built into the south wall. This clock was said to have included a diagram of the night sky, but when the tower was turned into a Christian baptistry and later a place of worship for the dervishes, the mechanism disappeared.[4]

The eight winds, shown flying clockwise around the building, are Boreas, the north, the personification of a strong wind; Skiron, the northwest; Zephyros, the west, carrying a heap of flowers to denote his newfound benignity; Lips, the southwest; Notos, the south, representing the wet winds of winter; Euros, the southeast, shown as a winter gale; Apeliotes, the east, carrying a basket of fruit; and Kaikias, the northeast, a strong wind; he is stocky, bearded, and looks none too friendly.

The tower was very precisely oriented. A thousand years before the compass, Andronicus plotted the directions exactly. Modern engineers wouldn't move the tower a third of a degree.

At about this time, the Romans began to weigh into the debate. Pliny the Elder, the Roman naturalist who was active a few decades after the start of the Christian era, got the origin of winds more or less right when he wrote in book 1 of his *Natural History of the World* that "the sun's rays scorch and strike everywhere on earth in the middle of the universe and, broken, bounce back and take with them all they have drunk. Steam falls from on high and again returns on high, empty winds violently swoop down and go back with their plunder . . . the earth pours breath back to the sky as if it were a vacuum." He wasn't very consistent (he asserted in book 2 that winds can just as easily "issue from certain caves in Dalmatia") but then he added, somewhat redeeming his naturalist's credentials, "neither is it impossible, but that they do arise out of waters, breathing and sending out an air, which neither can thicken into a mist, nor gather into clouds: also they may be driven by the lugitation and impulsion of the Sun, because the wind is conceived to be nought else but the fluctuation and waving of the air."

The fluctuation and waving of the air! Precisely. A modern meteorologist could hardly have put it more succinctly.

More winds exist than the four Homer described, Pliny wrote, but "the Age ensuing, added eight more; and they were on the other side in their conceit too subtle and concise. The modern sailors of late days found out a mean between both: and they put unto that short number of the first, four winds and no more, which they took out of the later. Therefore every quarter of the heaven hath [just] two winds apiece," which pretty much restated what the Horologium had been saying for a hundred years already. This assertion came in the middle of a rant against the sad materialism of modern times, in which the naturalist lamented that "men's manners are waxen old and decay; now, all good customs are in the wane: and notwithstanding that the fruit of learning be as great as ever it was, and the recompence as liberal, yet men are become idle in this behalf. The seas are open to all, an infinite multitude of Sailors have discovered all coasts whatsoever, they sail through and arrive familiarly at every shore: all for gain and lucre, but none for knowledge and cunning."[5]

That's pretty much where it rested for the next thousand years or so. Greek and Roman sailors plied the Mediterranean, Arab sailors the Indian Ocean, Chinese mariners the Yellow Sea, and the Phoenicians ventured out into the Atlantic and Indian oceans. Sailors are practical men, and they learned how winds can become a network of conduits taking them across the seas and back again; they understood that a typical journey might come to use a number of different winds to proceed in different directions—if one chooses carefully, one can always come home. As Sebastian Smith put it in *Southern Winds*, this was wind hopping, similar to changing buses several times to cross a city.[6]

Not just directions but also seasons were plotted, and dangerous winds were known to arrive at certain times of the year. Prudent rulers prohibited travel during those seasons to minimize their losses. In early Christian times the Coptic calendar listed precise dates for the beginning of bad winds. March 20 you could expect a two-day easterly gale, April 29 another. The *khamsin* was supposed to blow from the day after Easter to Pentecost. Ocean travel in the east Mediterranean was discouraged from St. Dmitri's Day (October 26) to St. George's Day (May 5).[7]

Over the centuries, the wind rose was developed, first appearing on Portuguese and Spanish charts by the thirteenth century. Pliny notwithstanding, early wind roses denoted the direction of thirty-two winds, eight major winds, eight half winds, and sixteen quarter winds. Diagrammed into a circle, these thirty-two points resembled the European wild rose, with its thirty-two petals, hence the name. In after years, the symbol of the rose itself became a beacon for the lost, and when the compass was invented, the wind rose mutated into the equally ornate compass rose, which, often embellished with puffing wind gods, was still shown on maps in the nineteenth century and is still occasionally added by cartographers to give a satisfactorily antique tone to their products.

A typical wind rose. Sometimes these have fanciful illustrations of puffing wind gods, often not. Sometimes the four principal Greek winds are named; in most cases they are taken for granted.

II

For the next few hundred years the two great branches of human endeavor—the practical, or artisanal, what we could today call the engineer; and the scholarly, or philosophical—went their separate ways, each developing its own brand of expertise. Practical men didn't bother with theory, the philosophers with experiment or observation. And so the theories of weather and wind that have survived tend to be the fruit either of pure reason innocent of observation or of visions; entirely missing are the thoughts of, say, a miller at his windmill, or a sea captain running before a gale, or a farmer who watched winds destroy or nurture his crops, or a roofer whose rafters collapsed in a storm. Typical of what did survive were the notions of eminent medieval scholar Hildegard of Bingen, a Benedictine sister at a nunnery in the Hunsruche near the Rhine. Sometime around 1145 she had a vision that the winds held all the elements together, each wind a wing of God working to keep the firmament in the right place, and causing it to rotate around the earth from east to west. Onto this lovely notion she then cobbled an impressively abstract view of how nature worked: The air was made of four layers, each governed by one of the cardinal winds. The east wind was closest to the ground; above it was the west wind, then the north, and finally the south. Within these layers everything else is to be found—the sun, the moon, all the constellations, storm clouds, and thunder. Hildegard was later beatified, but not for her work on the weather.

Throughout the medieval period, earlier theories persisted. Gervase of Tilbury wrote in *Liber de mirabilibus mundi* that "mountains and water cause winds." William of Conches maintained that four great ocean currents created the four cardinal winds. Sometime in the twelfth century Adelard of Bath produced his *Quaestiones Naturalies*, a compilation of seventy-six discussions of nature, including the weather. He ignored the Greeks, relying instead on imported

Arab science, then the most mathematically inclined culture on earth. "Wind," Adelard asserted, "is merely a species of air."

Even by the time Columbus sailed the ocean blue the split between philosophy and science was still evident. The effects of winds, and their general direction, were known. The existence of the trade winds was known. Sails and sailing vessels were becoming ever more sophisticated devices for actually employing the winds. But the science of meteorology had scarcely improved, and natural philosophers, as they were coming to be called, were still enjoined in arcane debates about exhalations from the earth or the sea.

Nevertheless, the warning signs of bad weather were understood by those who needed to know them. Sullen swells in an eerie calm meant a storm was due. Red sky in the morning, sailor's warning; red sky at night, sailor's delight—these things were generally known to be true. Columbus himself, who had suffered through a Caribbean hurricane on his first voyage, knew when one was due on a subsequent voyage. But he didn't know *why*. Or *how*. Above all, no one understood that weather *traveled*. Even the most apocalyptic storms were thought to develop, wreak havoc, and then dissipate, in one place.

Reconciliation of the two branches of knowledge was not to come for several more centuries. In 1582 the great astronomer Tycho Brahe began the work of systematization; he kept a meteorological daybook and began defining the winds not only by the direction but also by their force. His gradations included dead calm, two categories of light winds, five categories of strong winds, three of storms; it was the first wind scale, several hundred years before Beaufort's. But Brahe never took it that one step further, by actually measuring the speed of those winds. At the time, he had no way of doing so.[8]

In 1622 Francis Bacon, given lots of time for thought during a stint in the Tower of London, produced his *Historia Ventorum*, or History of the Winds. He still believed that wind was generated by vapors expanding their volume suddenly. Where the winds or the vapors actually came from remained opaque: "The places where there are great stores of vapours [are] the native Countrie of the Windes."[9] Ironically, Francis Bacon's illustrious ancestor Roger Bacon, the great medievalist and scientific prodigy, was closer to the mark more than four hundred years earlier when he simply noted in a journal that "heat makes air move." At the time, the observation went unheeded, mostly because the earlier Bacon was so prolix with his supply of ideas and inventions that his contemporaries were somewhat overwhelmed: He was the first person in the west, for example, to describe gunpowder; he invented spectacles for the eyes; and he was the first person anywhere, as far as is known, to propose mechanically propelled ships and carriages and an airplane with flapping wings. He was later imprisoned for his vocal contempt for his superiors in the Orders of Friars Minor and for his sarcastic views on the "puerilities" of other philosophers of his time.

Those same puerilities were still occasionally being trotted out even after the later Bacon. In 1668 Margaret Lucas Cavendish, the Duchess of Newcastle, published her *Grounds of Natural Philosophy* in which she asserted that "the strongest winds are made of the grossest vapours. Concerning the Figurative Motions of Vapour and Smoak, they are circles; but of Winds, they are broken Parts of Circular Vapours: for, when the Vaporous Circle is extended beyond its Nature of Vapour, the circumference of the Circle breaks into perturbed Parts; and if the Parts be small, the wind is, in our perception, sharp, pricking and piercing; but, if the Parts are not so small, then the wind is strong and pressing." As much "smoak" as this is, it did contain one new approach: an observation of what the winds actually felt like. Twenty or so years later, in 1684, Dr. Martin Lister, writing in the journal *Philosophical Transactions*, suggested that trade winds were caused by the constant breath of seaweed. In his view, their very regularity made their origin obvious: "The matter of that [ocean] wind, coming (as we suppose) from the breath of only one Plant it must needs make it constant and uniform: Whereas the great variety of plants and trees at land must needs furnish a confused matter of Wind."[10]

Despite these eminences, things had begun to change. Galileo helped. A proper lens-grinding technology helped. Precision instruments in the laboratory helped. The slow transformation of alchemy into chemistry helped. Artisans became educated and philosophers descended into the workshop. As early as 1627, the German Joseph Furtenbach fired a cannonball straight into the air to prove that the earth rotated. The ball, when it landed, was a little to the west of where it would have landed on an unmoving earth, no doubt to Furtenbach's relief (he was standing next to the cannon all the time). In 1639 Galileo essayed a variant of Aristotle's leather bag experiment. He manufactured a glass bulb with an airtight valve and sucked the air out. Then he weighed it. Next he forced air back in and weighed it again. There was a measurable difference. Aristotle's idea had been right, but his instruments defective, and Galileo was able to prove that air did, despite Aristotle's claim, have weight. A mere five years later, in 1644, Galileo's apprentice, Evangelista Torricelli, used his master's experiment to construct the first barometer, although the word *barometer* wasn't used until 1668, when Robert Boyle coined it for his own similar device. Whatever it was called, Torricelli's device was nevertheless the crucial breakthrough. For the first time a way existed to accurately measure a meteorological phenomenon.[11]

The practical men, too, were observing nature ever more accurately. In 1626 Captain John Smith, notorious exaggerator of his own derring-do ("puller of the longbow," as historian Samuel Eliot Morison put it) and best known for his role in the Pocahontas saga, put together his *Sea Grammar*, with its careful observation of the various winds and of the proper terms therefor: "When there is not a breath of wind stirring, it is a calme or a starke calme. A Breze is a wind blows out of the Sea, and commonly in faire weather beginneth about nine on the morning, and lasteth til neere night; so likewise all night it is from the shore . . . A fresh Gale is that doth presently blow after a Calme, when the wind beginneth to quicken or blow. A faire Loome Gale is the best to sail in, because the Sea goeth not high, and we beare out all our Sailes. A stiffe Gale is so much wind as our top-sailes can endure to beare . . . It over blowed when we can beare no top-sails. A flaw of wind is a Gust which is very violent upon a sudden, but quickly endeth . . . A storme is knowne to every one not to bee much lesse than a tempest, that will blow down houses, and trees up by the roots . . . A Hericano is so violent in the West Indies, it will continue three, foure, or five weekes, but they have it not past once in five, six or seven yeeres; but then it is with such extremity that the Sea flies like raine, and the waves so high, they over flow the low grounds by the Sea, in so much, that ships have been driven over tops of high trees there growing, many leagues into the land, and there left." A little longbow-pulling there in his description of a *hericano*, but otherwise useful enough.

In 1654 John Winthrop, the first governor of Massachusetts colony, recorded the first use of the current spelling of *hurricane*, when he wrote in his *History of New England* of the "great colonial hurricane" that had just passed by.

In 1663 Robert Hooke suggested "a Method for making a History of the Weather" by observing and recording "the Strength and quarter of the Winds." He recommended a scale from one to four, which included half numbers, so it was really a scale of nine.[12]

Eminent explorer and notorious pirate William Dampier was even more precise. In 1687 he encountered a major storm in the South China Sea: "Typhoons," he wrote, "are sorts of violent whirlwinds"—the first time this observation had been recorded. "Before these whirlwinds come on . . . there appears a heavy cloud to the northeast which is very black towards the horizon, but towards the upper part is a dull reddish color. The tempest came with great violence, but after a while the winds ceased all at once and a calm succeeded. This lasted . . . an hour more or less, then the gales were turned around, flowing with great fury from the southwest."[13]

In the middle of the eighteenth century, astronomer Edmund Halley, of comet fame, published an article in *Philosophical Transactions* that declared the cause of winds to be the heating of air by the sun. He wasn't quite right—he suggested that winds blow primarily from the east because the sun rises there, thereby making the classical mistake of generalizing from an untypical particular, the fact that at his home the morning winds were easterlies—but his article did suggest that the rotation of the earth had an effect on weather. And the article was memorable for another innovation: He published the first crude weather map.[14]

After that, developments came thick and fast. Anders Celsius, who constructed a temperature scale, also put together a wind-force scale. In Germany the Palatine Society, which joined the mania for measuring winds, set up the first weather office by coordinating observations from several cities in the Mannheim region of Germany.[15] The ever-curious Benjamin Franklin added to the storehouse of knowledge in 1743, after a storm blocked a sighting of an eclipse of the moon he had been particularly eager to see. The winds in Philadelphia were from the northeast, but he discovered through correspondence that Boston, which was to his northeast, actually suffered the same storm later than he had. A few years later this curiosity led him to a theory of storms, and he concluded they were disturbances that moved independently. He didn't know yet that they rotated. If he had seen Dampier's observations, they hadn't registered with him.[16]

By the end of the century the practical men understood a good deal about storms, though a more careful calibration and a really comprehensive theory wasn't to be for another few decades. Sea captains understood clearly that pressure affected weather—that low pressure meant a storm—and no ship would leave port without a glass, as barometers were by then called. By the early nineteenth century meteorologists were commonly drawing isobars on maps to denote pressure, which meant winds.[17] In 1802 Nathaniel Bowditch produced his *American Practical Navigator*, which was filled not only with graphs and charts (one of his graphs suggested that a drop of one millibar in an hour indicated a storm center twenty-two miles away; a drop of three millibars a storm center ten miles away) but with hard advice to sailors that made it clear just how storms actually traveled, and what they could do about it. A storm's center could be found by the simple expedient of facing the wind, in which case the center would be to your right. If you did this periodically, you could tell which way the storm was moving. The worst possible place to be was in the direct path of the storm's advance or toward its right—that is, to the north of a storm tracking west, to the east of a storm tracking north. Bowditch called that the dangerous semicircle, where the winds actually pushed vessels back into the center of the storm. The safest place, by contrast, was to the left of the path, where the winds would tend to propel you away from the center—and away from the center was very definitely where you wanted to be. He called this the navigable semicircle.[18] All this made sense only in the northern hemisphere, because the storms were rotating counterclockwise.

Later Joseph Conrad had some fun with these rather head-scratching directions, and in *Typhoon* had his captain mulling the possibilities: "[The skipper] lost himself amongst advancing semi circles, left and right hand quadrants, the curves of the tracks, the probable bearing of the center, the shifts of wind and the readings of the barometer. He tried to bring all these things into a definite relation to himself, and ended by becoming contemptuously angry with such a lot of words and with so much advice, all head work and supposition, without a glimmer of certitude. 'It's the damndest thing, Jukes,' he said. 'If a fellow was to believe all that's in there, he would be running most of his time all over the sea to get behind the weather.' "[19]

Finally, in 1831, the true cyclonic nature of storms was persuasively described by the autodidact scientist William Redfield. His conclusions were drawn from careful observation, especially by tracking detailed reports of hurricane damage after a storm had passed through Connecticut. He learned from these reports that trees had been felled in different directions, depending on where they had been in the storm's path. And it became clear that although at certain places the winds had been driving in from the north, those same northerly winds felled trees in southerly counties before they did their damage in the more northerly ones. That was Redfield's major eureka moment. The only explanation possible was that the storm was, in essence, a giant whirlwind. His conclusions, called "On the Prevailing Storms of the Atlantic Coast," were published in the *American Journal of Science*. He didn't call these storms *cyclones*—that term, derived from the Greek word meaning a coiled snake, was invented some time later by an Englishman, Henry Piddington, who applied Redfield's data to the massive storms in the Bay of Bengal. Nevertheless, the circular nature of the storms was conclusively established. Redfield's data also showed that cyclonic winds move in spirals, and not in concentric circles, that the velocity of rotation— the sustained winds—increases from the fringes toward the center, and that at the same time the whole body of the storm is itself moving, at a speed far less than that of the rotational winds.[20]

Many other scientists, such as Elias Loomis of Cleveland and James Pollard Espy of Philadelphia, later confirmed Redfield's data. Espy, who in 1842 was a professor at the Franklin Institute, became the first official U.S. meteorologist. It was Espy who contributed the final piece of the puzzle of great storms: the notion of rising air and latent heat. He proved that when moist, warm, rising air cools and precipitates out, it releases heat, for the curious but simple reason that molecules of water contain less energy than molecules of vapor. It is this latent heat that reheats the cooling air, causing it to rise farther, thereby drawing more air up after it; it is the key to the self-sustaining nature of tropical cyclones, and an explanation of their awesome power—they are self-generating furnaces and continue to exist as long as a supply of fuel, warm and moist air, can be found at the surface.

The practical men and the natural philosophers had at last come together, being described for the first time in the nineteenth century by the new word *scientist*. One of these new scientists was Matthew Fontaine Maury, born in 1806 in Virginia, who joined the navy and within a few years had made three voyages, to Europe, around the world, and to the Pacific coast of South America. He then spent the years 1834 to 1841 producing voluminous works on sea navigation and plotting the best paths for sea voyages. His best known work, *Explanations and Sailing Directions to Accompany the Wind and Current Charts*, contained chapters on the atmosphere, on "red fogs and sea dust," on the winds, and on matters as diverse as the equatorial cloud ring, the salts of the sea, the ocean currents, the Gulf Stream, the influence of the currents on climate, the depths of the ocean, the Atlantic Basin, and on gales, typhoons, and tornadoes. In 1842 he was appointed superintendent of the Depot of Charts and In-struments for the U.S. Navy, where he developed a system for recording the oceanographic and atmospheric data provided by both naval and merchant marine captains, and published the results in 1855, in his *The Physical Geography of the Sea*, the first real text-book of modern oceanography. One sea captain reported that he had followed Maury's instructions and had cut the duration of a voyage from New York to Rio from forty-one to twenty-four days, and it wasn't long before marine merchants insisted that their skip-pers use the new scientific navigation techniques too. Maury re-signed from the U.S. Navy when Virginia seceded from the Union, and became commander of the navy of the Confederate States.

In a curious byway of science history, Maury has more recently been adopted by the woollier fringes of the Christian far right, who have come to believe, erroneously, that he was prompted to discover the Gulf Stream and other ocean currents through interpretation of a biblical passage on the "paths of the sea." In fact Ponce de León had written about the Florida current in the early 1500s, and a chart by Benjamin Franklin, published in 1786, well before Maury's birth, clearly shows the Gulf Stream.

Contemporaneous with Maury was the work of William Ferrel,

whose *Essay on the Winds and Currents of the Ocean* rediscovered the forgotten work of Gustave-Gaspard Coriolis. It was in this essay that Ferrel, a self-taught farm boy from what is now West Virginia, proposed his famous model for the midlatitude circulation of the earth's atmosphere that was late to be called the Ferrel cell. His theory was that air flows toward the pole and eastward near the earth's surface, and toward the equator and westward at higher altitudes. His theory doesn't match precisely what actually happens, but it was still the first real explanation for the westerly winds in the middle latitudes of both hemispheres.

Only after manned flight in the twentieth century were the overall patterns of air circulation finally plotted. The work was given some urgency in the First World War, because the commanders of the new air forces desperately needed data they could use to protect their lethal but nonetheless fragile little bombers. By the late 1920s it was understood that winds were the continuing collision of huge air masses in waves, fronts, ridges, and troughs, all caused by solar radiation and the rotation of the planet. The final piece of the puzzle—the discovery of the high-altitude stratospheric winds and the jet streams—had to wait until aircraft could fly higher still. By the Second World War a real understanding of winds was, finally, in place.

CHAPTER FOUR

Wind's Intricate Patterns

*I*van's story: *By the afternoon of September 3
the system had moved almost 300 miles farther into the tropical Atlantic,
and was located 865 miles south southwest of the Cape Verde Islands.
Around it and above it the air was still, but the system itself was still spinning slowly, moving just south of west at about 30 miles an hour. Strong
winds stretched outward about 60 miles from the center. Within a few hours
the winds picked up and passed the 39-mile-an-hour threshold that turned
the system officially from a tropical depression into a tropical storm. A bulletin from the National Hurricane Center in Miami assigned the system the
name Ivan, according to custom and its place in the annual cycle, and found
it was strengthening steadily. By five P.M. on the 3rd, Bulletin 5 found that
the storm's central pressure had dropped to 1,000 millibars, and maximum
sustained winds were in the region of 50 miles an hour, with higher gusts.
This was still well shy of the 74 miles an hour of a hurricane, but the bulletin was cautious: Strengthening was expected, and it could turn westward
onto a new track.*

*In the small print of the bulletin, issued under the signature of Forecaster
Beven, there were hints of what Miami thought would happen. "Thought,"
because they didn't know; winds are predictable in their larger patterns and
behaviors, but horribly intricate in their local behavior. Aircraft data had found
flight level winds of 95 knots near the center.[1] Beven expected the system to
turn slightly north and reach hurricane status within two days. Further ahead
than 36 hours, he wrote, the forecast was of low confidence.*

He expected Ivan would head for the Gulf of Mexico, but the system could still go anywhere, do anything.

That bulletin was the first time Ivan impinged on my consciousness. I logged on to the National Hurricane Center Web site and squinted along the predicted track. It looked likely that Ivan could go to Puerto Rico and then Cuba, and after that southern Florida and the Gulf. It didn't appear dangerous for the northeast corner of America. But that "low confidence" niggled. The northern edge of the track forecast would take the storm north of the Bahamas, where it could curve northward, as most storms do under the influence of the Coriolis force, and head for Bermuda. That could be Trouble.

At least for me. I confess I was rather less concerned at this point about Florida than I was about the storm showing up in the northern Atlantic and bashing away at my beach. I figured Florida was more able to cope with Trouble than I was. NIMBY, Not in My Back Yard, yes. Sorry.

I

Sometimes you really can see the wind. Directly, not just by watching trees sway or ripples skitter across open water. Last year we had a two-day blizzard that dropped almost a yard of snow in gusty 50-mile-an-hour winds. At its height, you could stare out the window and watch the white wind dipping and swirling through the trees, heaping the snow up in great sculptured drifts, darting around the corner of the house. We had often wondered why the greatest damage in wind storms seemed to be done to a row of rhododendrons on the lee side of the house, and suddenly we could see the reason for ourselves—the wind dashed itself at the corner of the building and split, some of it going straight up and tugging at the eaves but the rest swirling around the corner, where it was squeezed between the house and a cut in the bank, suddenly accelerating just before it passed the rhodos, tearing at their leaves. I had already "seen" the wind in the violent dust storms of my boyhood in South Africa, where the prevailing color was not white but a brutal violet, and later in Saharan sandstorms, where the wind blows a gritty ochre and dun, and of course tornadoes are lethally visible. But these are crude views, made visible only by wind's awful force. And so when I read a lyrical passage in Sebastian Smith's sailing memoir, *Southern Winds*, I found myself nodding in agreement: "Sometimes, staring into the sky, I tried to imagine what it would be like to see the wind. As if there might be special glasses to understand [the wind's] secrets—the paths used to reach seemingly impossible places, the mystery of opposing airs and the drama of the katabatic wind. In simulations these can be created but to ordinary men they remain as inscrutable as any god . . . Imagine standing on deck and being able to watch jet streams swoop through the summer skies. Or the slow turning wheel of low-pressure systems. Or the avalanche of a squall."[2] Just so. How wonderful to be able to watch the wind's intricate dance across the planetary surface and high into the troposphere, to see how great air masses move and collide and meld and wrestle, how little zephyrs tickle the fine hairs on the forearm, how the breezes tease their way through trees and over rocks and up hillsides. Sometimes airplanes hit an air pocket—which is really vertical-shear wind plunging violently downward through pressure differentials in the surrounding air. Wouldn't pilots, and their passengers, love to see that wind before they hit it?

We can understand wind—we understand it now almost to the molecular level—but it would be much more agreeable, and infinitely more useful, to be able to see it.

Wind starts, of course, as with so much else on our planet, with the sun.

Deep inside the seething cauldron of the sun, thousands of miles beneath the corona, or what passes for its surface, is a constant cascading series of hydrogen fusion reactions. Every nanosecond millions of hydrogen atoms crash together, and for every four that destroy themselves in this furious suicide, one helium atom is created. Since four hydrogen atoms weigh slightly less than one helium atom, a fraction of the mass is lost each time, and this shortfall—*pace* Einstein—is released as pure energy.[3] So much energy, indeed, that the temperature of the sun is maintained at a fairly steady 15 million degrees Celsius. Some of that energy is radiated into space. A tiny fraction, about two billionths, reaches the earth.[4] This might not seem a lot, but the sun is so massive relative to Earth that the solar radiation reaching us is around 175 trillion kilowatt hours of energy every hour.[5] Somewhere between 1 and 2 percent of this energy input is converted into wind, about fifty to one hundred times more than all the energy converted into biomass by all living things on Earth.

These trillions of watts of solar energy strike the earth directly at the equator and more obliquely closer to the poles. This was far from obvious in ancient days, when the earth was a flat disk and the sun directly overhead, but to us the mechanism is obvious—equator at high flame, midlatitudes at a simmer, poles just barely affected, a straightforward consequence of Earth's spherical shape, a simple pattern complicated only by Earth's rotation on its rather tilted axis and its annual rotation about the sun. Pretty obviously, the air is therefore hotter at the equator and cooler at the poles, and it is in these differentials that all winds, and all weather, and therefore climate, are derived.

The differentials are set in motion by one simple governing principle: entropy. Entropy, disorder—or "mixed-upness," as Richard Dawkins called it—is the substance of the second law of thermodynamics, which is that entropy, or disorder, always increases in a closed system. In this definition, order means that different parts of a system have different characteristics (heat, pressure, odor); disorder means no part is different from any other. In nature, if something is hot and something is cool, reordering will occur. That is, if a zone of high pressure is near a zone of low

pressure, nature will try to equalize the two zones through movement of air from the high to the low, and winds result. Nature is
striving for balance; climate—derived from the Greek word *klima*,
meaning degree of latitude—is the earth's way of seeking to balance its energy intake.

This is just as well. If this didn't happen—if entropy wasn't at
work, if there was no balancing—there would be no wind, no
weather, and no life on earth as we know it. The poles would go
into a much deeper freeze, the equatorial zones would overheat,
and what little organic life remained would be huddled at the in-
terstices.

And so, in some ways winds, the movement of the air relative to
the surface of the earth, are simplicity itself: Hot air one place,
cooler air another, and there you have it—wind. Pressure differentials—wind. Adjacent climate zones—wind. Altitude differentials—
wind. Planetary rotation—deflected wind. The physics is not
complicated: Wind is air moving from high pressure to low pressure, in a straight line, deflected by the rotation of the earth (the
Coriolis force).

Because winds begin with the sun, the key to understanding
global wind patterns is to start where solar radiation is most intense,
at the equator. The air warmed by the radiation rises quickly, causing a quasi-vacuum of low pressure that draws air toward the equator from semitropical latitudes. The winds so created head directly
for the equator but because of the planet's rotation are turned by
the same Coriolis force that twists the ocean's currents. This "turn"
pushes the winds "right" in the northern hemisphere and "left" in
the southern hemisphere, until they parallel the equator. These are the
most reliable winds on earth, the so-called trade winds; they are the
winds that made transoceanic travel possible in the days of sail.
Eventually these steady trade winds, because they are paralleling the
equator, are also warmed, and they then follow the same pattern—
they rise, are cooled, drift toward the poles, and sink again, at about

30 degrees of latitude, more or less the southern Mediterranean and northern California.

Some of this newly cooled air moves at high altitudes back toward the equator, completing what is known as the Hadley cell, named after George Hadley, an English lawyer in the eighteenth century. But some of it moves toward the midlatitudes, and on its way the Coriolis force "turns" the wind right in the northern hemisphere and left in the southern, causing the prevailing midlatitude westerlies, the winds that the canny New Englanders learned to exploit when their trade with Europe began to expand. But the westerlies only account for a fraction of the air mass in motion. The rest, which forms a second overturning cycle of air, carries surface air poleward and upper-level air back toward the Hadley cells. These are Ferrel cells, which generally have a motion opposite to planetary rotation; they exist largely to balance the Hadley cells and their equivalents at the poles.[6] Ferrel cells and Hadley cells meet at the horse latitudes.

This intricately three-dimensional pattern of winds is the planet's general circulation. It explains why most of the air movement in the lower atmosphere is vertical rather than horizontal. It is also directly responsible for the latitudinal movement of air masses, and therefore of weather and the long drifts of weather called climate.

II

This business of the Coriolis force, named after French physicist and mathematician Gustave Gaspard Coriolis, who first described it in 1835,[7] merits a small digression, because it is not quite as simple as it sounds. The physics that govern what scientists call a rotating frame of reference are quite complex, but the effects of the earth's rotation on natural phenomena are simple enough to see. East–west motions create a Coriolis force, or Coriolis effect, which is directed radially inward (for easterly motion) or outward (for westerly motion) from the axis of rotation. Motion relative to the earth's, to the

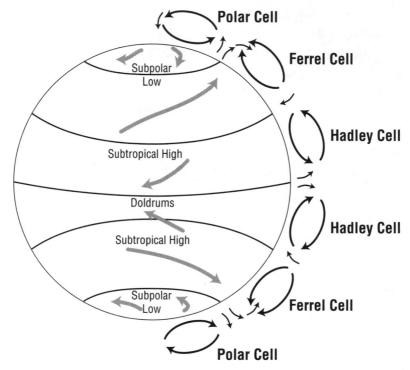

Hadley and Ferrel cells, showing a simplified version of the major vertical air movements that help balance planetary heat distribution.

west or the east, will produce an acceleration (because of the force) to the north or the south.

North or south motions also give rise to a Coriolis force because the motions are toward (or away) from the axis of rotation. Vertical motions also give rise to a horizontal Coriolis force, but it is negligible and usually ignored.

For our purposes, the result is what is important: Whether from east-west or north-south motions, there is always a deflection to the right in the northern hemisphere, and to the left in the southern hemisphere.

The easiest commonsensical way to see the effect is to imagine

Gustave Coriolis

firing a rocket or an artillery shell while you are standing at the equator. Aim the rocket to hit a target a thousand miles away. If you fire it toward the north pole, something odd seems to happen. Even if your calculations were spot on, and there was no wind, the rocket wouldn't land due north of where you are. Instead, it will appear to have drifted off course, to the east. Surely something must have diverted it? Was Newton wrong?

The answer lies in the rotation of the earth. All points on Earth rotate 360 degrees in twenty-four hours, a planetary day, but obviously some points must be rotating much faster than others. The fastest speeds are at the equator, perhaps a thousand miles an hour; close to the poles the rotational velocity is negligible, a mere dozen or so miles an hour.

So what has happened to your rocket is this: At the moment you fire it in a northward direction, you and the launcher that fired it,

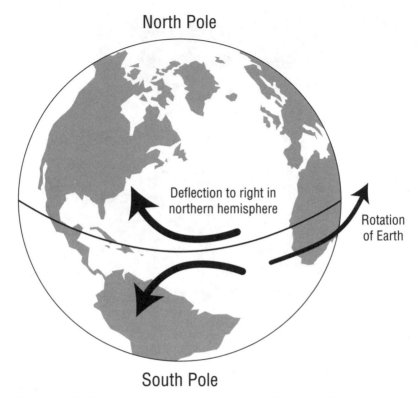

North Pole

Deflection to right in
northern hemisphere

Rotation
of Earth

South Pole

The Coriolis force, showing how movement is deflected to the right in the
northern hemisphere and to the left in the southern hemisphere.

and therefore the rocket itself, are traveling eastward at a high rate of
speed—the speed the equator is traveling. Your rocket travels in a
straight line, but throughout its flight it keeps moving eastward at a
constant rate, the speed of the equator. When it nears its destination, however, the ground below it is not moving eastward nearly as
fast as the rocket itself. And so the ground "seems" to have moved
westward, and the missile "seems" to have drifted eastward. That is,
an object traveling away from the equator will eventually be heading east faster than the ground below it, and may seem to have been
driven east by some unknown force. Objects traveling toward the

equator, similarly, will seem to be have been driven west. Which means, if you can turn your head to squint in the right direction, that in the northern hemisphere objects will turn to the right, and in the southern hemisphere to the left.

Newton's law is conserved, after all. It is just that we are in a rotating frame of reference.

Some of the effects are critical ones, and not just for long-range ballistic missiles. Over the course of a four-hour flight, a jetliner traveling north or south must compensate for the Coriolis effect, or a pilot from Toronto will find himself depositing his passengers in New Orleans instead of Atlanta.

On a very small scale, say that of the average bathtub or kitchen sink, the angle of deflection is so tiny as to be unnoticeable, estimated at 1/3600th of a degree, so the way water flows out of a bathtub depends on the angle it was poured in, the configuration of the tub itself, abrasions on the surface, the temperature differentials in the water, and dozens of other minute effects. It is theoretically possible, in a carefully controlled laboratory experiment, to show the effect at work; most of us, to see it at all, would need a tub the size of one of the Great Lakes.[8]

I once saw a "demonstration" of the Coriolis force by a cheerful Maasai at the Equator near the town of Eldoret, in Kenya. I had come to Eldoret after a rather harrowing all-night, breakneck drive in an aging Renault, piloted by a former game warden; we shared the car with a massive python, which the game warden called Brenda, that creaked and rustled around in the backseat and kept me awake. I was therefore feeling somewhat fragile when the car stopped at the "Line," where we joined a busload of tourists who were having their pictures taken with one leg in each hemisphere, and who were being subjected to the aforementioned Coriolis demo. Two buckets had been filled with water and placed a yard on either side of the line. Then the demonstrator, with a flourish, pulled the plug on each in turn, and lo and behold, the water exited by spinning counter clockwise in the northern hemisphere bucket, and clockwise in the southern. This produced the requisite "ooohs" and "aaahs" from the busload, as well as a satisfactory yield in tips, but it was all a fraud, set up by the direction in which the con artist had poured the water into the buckets, and by a minute difference in the way he pulled the plugs. The game warden, whose name was Ibrahim, gave him a disgusted snort for his pains, and so did I, but he wasn't offended; he was making a good living from human gullibility and the occasional skeptic wasn't going to deter him.

The Coriolis effect influences massive movements of air just as much as it does artillery shells or aircraft, and is therefore important for an understanding of global winds. Just like an artillery shell, freely moving air (winds) will deflect to the right in the northern hemisphere, and to the left in the southern. So air moving toward a low-pressure system will deflect to its right; but because the forces which got the air moving toward the north in the first place are still in play, the result will be a vortex of air, spinning counterclockwise—air will try to turn to the right, the low pressure pocket will try to draw the air into itself, and the result is that the air is held into a circle that actually turns to the left. Without the Coriolis effect, air rushing into a point could still form a vortex, but the direction of rotation would be random.

With the Coriolis effect in play, the randomness disappears, and northern hemisphere cyclones, including hurricanes, always revolve in the same direction.

Bear in mind that the Coriolis force is not the only factor determining large scale winds. One force is generated by the pressure gradient, of course, but there are three others: the curvature of the wind (wind that turns speeds up or slows down depending on whether it is turning clockwise or counterclockwise [the opposite in the southern hemisphere]); changing pressure differences (called the isallobaric effect), which can dramatically boost or inhibit the wind, a major consideration in east coast "weather bombs"; and the

stability of the air in the lowest part of the atmosphere (stable air, either warm or cold, is reluctant to overturn, whereas unstable air is
highly turbulent, and its overturning and mixing brings down the
established winds from higher elevations to where we live).

III

The general circulation model—the Hadley and Ferrel cells and the
generally vertical overturnings of huge air masses—is important for
understanding global winds, but since most of us see winds as horizontal, not vertical, it's more useful too look at planetary winds in
another way: at how solar radiation and the Coriolis force together
conjure into being a consistent pattern of latitudinal "belts" paralleling the equator.

Starting from the equator and going north, these belts are the
doldrums, the trade winds, the horse latitudes, the prevailing southwesterlies of midlatitudes, and the northeasterlies of high polar latitudes. In the southern hemisphere, the same belts exist but the wind
directions differ.

The doldrums straddle the equator and girdle the earth, a belt of
low pressure, static and ever-present, a windless region known to all
sailors. The doldrums are more properly known as the intertropical
convergence zone (ICZ), or thermal equator, which generally occu-
pies about 5 degrees each side of the equator, though it migrates
somewhat north or south with the seasonal position of the sun. (It
also shifts farther to the south over land masses such as South America, Africa, and Australia, and farther to the north over open water,
such as the Pacific or Atlantic.[9] It can occasionally reach beyond the
30th parallel.) Confusingly, it is sometimes called the equatorial
convergence zone or the intertropical front.

The trade winds are next, bounded on the doldrums side by a
zone of sharply rising winds, creating towering cumulonimbus
thunderclouds and torrential rains. The trade winds flow from the

next band out, the subtropical high-pressure zone called the horse latitudes, toward the low-pressure zone of the doldrums, and are "turned" westward by the Coriolis force. They were named, obviously enough, for their useful ability to push sailing vessels quickly and economically across oceans; they blow steadily at about 12 miles an hour between the ICZ and the 30th degree of latitude. In the northern hemisphere, the trade winds are northeasterlies; in the southern hemisphere, they are southeasterlies.

The horse latitudes are bands of calmer air, zones of stillness that have caused many a mariner to rue his profession. Maritime lore has several explanations for the name, all of them more or less implausible; the most popular is that ships became becalmed for so long in this zone that sailors and colonists were forced to toss their horses overboard, no longer having enough water to keep them alive. The horse latitudes are sometimes called the calms of Cancer in the northern hemisphere and the calms of Capricorn in the southern hemisphere. On land, most of the world's great deserts lie in this region.

Outside the horse latitudes are the prevailing westerlies (southwesterly in the northern hemisphere, northwesterly in the southern hemisphere) that cover most of Europe, North America, China, and similar latitudes south of the equator. They're as consistent—and as useful—as the trade winds. In the southern hemisphere they're closer to the equator and are more vigorous than their counterparts in the north, and are known as the roaring forties. In the northern hemisphere, this is where most of the North American and European weather is generated.

In high latitudes the flow reverses, and easterlies are the dominant wind patterns of both Arctic and Antarctic regions.

The most turbulent, changeable and perversely complicated weather on the planet occurs in the midlatitudes, where the warm equatorial air and the cooler polar air intersect in an apparently patternless turbulence. This is the downside of living in temperate

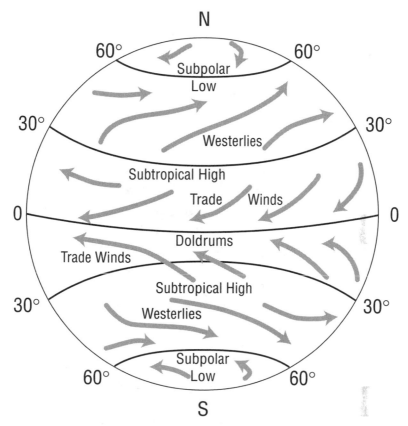

The prevailing winds, showing the three major bands of steady planetary winds: the trade winds, the midlatitude westerlies, and the subpolar easterlies.

regions, otherwise the most benign climate on Earth. The intersections of these winds—frontal zones—cause violent storms, tornadoes, thunder cells, and gales, as well as mild breezes and balmy temperatures. Mark Twain was moved to comment on the weather in northeastern America: "The people of New England are by nature patient and forbearing," he said in a speech to the New England Society in 1876, "but there are some things which they will not stand. Every year they kill a lot of poets for writing about 'Beautiful

Spring.' These are generally casual visitors, who bring their notions of spring from somewhere else, and cannot, of course, know how the natives feel . . ."

North Atlantic weather is complicated further by more or less permanent zones of low pressure around the 60 degree mark, the latitude of Greenland, and zones of high pressure in the horse latitudes, around the arid zones of the Middle East and the American southwestern deserts. In the Atlantic, weather forecasters call this the Bermuda high, and its presence, strength or weakness, is crucial for forecasting Atlantic weather; it can help "steer" hurricanes and other low-pressure systems. This high migrates east and west with varying central pressure. In summer and fall it is located near Bermuda; in the winter and early spring it is primarily centered near the Azores Islands, and then—surprise!—it is called the Azores high. The actual flow depends on a number of incalculables, including seasonality and jet-stream positioning. The circulating air tends to skirt the highs and get funneled into the lows, a tendency that is further complicated by topography (air flows more smoothly over water than over land, and less smoothly over mountains than plains), and even further complicated by the smoothly flowing stratospheric winds, that sometimes have the benign effect of shearing the tops off massive cyclones before they can coalesce into hurricanes.

Tropical cyclones such as hurricanes or typhoons usually decay as they reach cooler higher latitudes. But sometimes they decay in dangerous ways, interacting with the unstable baroclinic environment in higher latitudes and turning into what are called extratropical cyclones, a process known as extratropical transition, or ET for short. ET events, as they are called, can reenergize waning hurricanes and carry massive amounts of moisture, with the risk of flooding on land and storm surges along the shores. By some measures, almost half of the tropical cyclones that form in the Atlantic undergo some form of ET. New Zealand gets a few of these in the southern hemisphere, but the global champion in ET events, the one place where they are more likely to occur than any other, is northern New England and Atlantic Canada. One of the tasks of the Canadian Hurricane Centre has been to better understand ET; in the year 2000, Chris Fogarty, a meteorologist who operates out of the center, and others flew into Hurricane Michael while it was undergoing transition, and reported on the results in the journal of the American Meteorological Society. At the very least, they concluded, there is an urgent need to develop numerical and conceptual models that will enable weather forecasters to better anticipate changes and to improve warnings.[10]

A "baroclinic" storm enveloped us in the winter of 2004. On the morning of February 17, 2004, the forecasters told us to expect snow. By midday, the forecast had changed to "expect heavy snow," and by evening a blizzard warning was in effect. They had spotted a low-pressure system that had organized southeast of Cape Hatteras, in North Carolina; it had taken on the classic cyclonic shape as cold air from eastern North America clashed with the warm waters of the Gulf Stream, and was heading northeast at almost 24 miles an hour. By the morning of the eighteenth, there was a broad cloud shield and a high cirrus deck spreading out over the region, the storm already beginning to stretch out to the northeast in a comma shape. A strikingly sharp cold front extended southward as far as the Bahamas. The barometer was dropping fast, and in just over twenty-four hours had gone from 1,000 millibars to 959.

The forecast did its job. Everyone in the storm's path knew it was coming. Snowplow crews gassed up, power company workers checked their equipment and cranked up the ATVs. We stocked up on drinking water in case the power went off, and inventoried our food supplies; we made sure there was enough firewood for the fireplace and checked the kerosene supply for the lamps—we were reasonably well prepared.

We woke up to a howling wind and heavy snow. It snowed all

day. It blew all day. It snowed all night. It blew all night. Shutters banged, trees cracked, the entire landscape seemed to groan in pain. Below the irritant of the banging shutters there was a constant basso roar, as the wind-born waves hurled themselves at our rocky beach. I wasn't worried they'd come over the rampart—only the rare hurricane causes that—but it still sounded like an express train going by at high speed. I seemed to feel the bedrock shaking, though I knew it was my imagination. Out on Sable Island, which was nothing but sand, the giant waves literally did make the whole island quiver, but here . . .

On top of the roar of the waves and the clattering shutter was the sound of the wind gusting in the spruces, varying in pitch from a howl to a whine. On the morning of the nineteenth, it was still snowing, the wind still blowing, visibility only a few yards. The storm had slowed down. It was not quite stalled, but it was now centered over Sable Island, a hundred miles to our east, and was drifting in a leisurely way to the northeast. We had maybe three feet of snow on the ground, swirled around and piled high in drifts. I couldn't see the cars, except for their antennas. The front door had snow piled against it well over the door handle. The back door, in a wind shadow, was completely clear, but halfway up the path the snow was over my head. Our deck was buried in a snow cover maybe four feet thick. The picnic table had disappeared entirely.

The media in the next few days took to calling the storm White Juan, in memory of Hurricane Juan, which had passed through only five months earlier. Winds during the storm were steady at 48 miles an hour, with gusts up to 75 miles an hour in exposed areas, decently into the hurricane range.[11]

An intense storm, but not that atypical in winter. Perhaps more snow than usual, but every year in these latitudes there are two or three storms pretty much like it.

What's really interesting about these hurricane-force winter gales is not their animus but the mundane and mechanically simple nature of their origins. The western edge of the Gulf Stream is where these storms are made. As the current snakes its way past Cape Hatteras and turns back out to sea after a close swipe at land, it mingles briefly with the cold tongues of the southbound Labrador Current. In winter, when a dome of high pressure from the Arctic drifts southeast, bringing with it vast eddies of freezing air, it may come to the edge of the Gulf Stream and stall. If a core of warm air is encountered off the coast, just east of the Gulf Stream, and if the jet stream is flowing toward the northeast, the two air masses collide, the cold air fighting to move east, the warm air prodded by the jet stream. A pocket of turbulence develops in the crook between them. Wind flows east, then is bent quickly to the north. Unable to resist the centrifugal force, it begins to move full circle, creating a system of low pressure that deepens violently.[12]

Because the winds are flowing counterclockwise around these storms, the winds come out of the northeast as the storms move offshore, and the fishermen just call them nor'easters. So commonplace are these violent winter storms that the load lines painted on modern ocean vessels denoting the depths to which they may safely be loaded always have as the lowest line, the lightest loading, a line marked "WNA"—Winter North Atlantic.[13] Storms strong enough to severely damage large vessels happen on average about once or twice a year, particularly in the wintertime.[14] Such storms happen in all the midlatitudes of the northern hemisphere. Insurance companies have tracked the damage they do; the worst in the last one hundred years was a 1953 storm that made its way to Europe, causing massive Atlantic storm surges, killing almost two thousand people. (For some of the worst winter storms, see Appendix 8.)

Exactly why some mixes of cold and warm air deepen explosively and others remain benign is still unknown. It all depends, says

Peter Bowyer of the Canadian Hurricane Centre, on how fast the cold dry air and the warm moist air masses are forced together. What happens after they deepen is much better understood. Bowyer tells me, "Our computer models are quite adept at handling these storms. The physics of winter storms is better understood than that of hurricanes, so our models do quite well, and we can predict many days ahead of time when a storm is likely to form over, say, the Carolinas, and how long it will likely take to reach our region."

High overhead in both hemispheres are the stratospheric gales, and the jet stream of popular weather forecasting. This jet stream, which is essentially a standing wave of high pressure pushed eastward by the Coriolis force at the boundary between cool polar air and warm tropical air, flows at speeds up to 240 miles an hour, and sometimes more, at elevations around 30,000 to 35,000 feet, a little over five miles. Jet streams are generally the reason why long-distance flights are faster going eastward, because airline bosses like pilots to ride the jet stream to conserve fuel. Not that this is always easy; jet streams do meander, and substantial vertical wind shear, which can cause clear-air turbulence, can often be encountered at their edges; passengers who'd like a quick crossing don't necessarily sign on to a roller coaster and, as cabin crews can attest, are easily irritated by bumpy rides. Jet streams were first discovered during the Second World War, which was when high-altitude transatlantic flying first became commonplace. They were dubbed jet streams because they seemed to flow in narrow ribbons at high speeds—and jet aircraft had just been invented. More than one jet stream exists—the midlatitudes jet stream, the polar jet stream, and the polar vortex—and they are found above all oceans and all continents. The first to exploit them were the Japanese military, which in 1944 and 1945 launched experimental bombs suspended from balloons into the Pacific midlevel jet stream; some of these weapons traveled five thousand miles in three days; one reached the coast of Oregon, exploding near a Sunday school picnic, killing five children and the minister's wife.[15] Another made it over the Rockies and reached the thinly populated Canadian prairie province of Saskatchewan, a most unexpected assault.

A jet stream, too, can be deflected by pressure systems and is generally closely watched by meteorologists in winter—a jet stream represents an area the scientists call a zone of baroclinic instability, and a deflected jet stream in midlatitudes can mean the difference between mild temperatures and severe ones.[16] Jet streams that dive southward into the United States typically mean intense cold over much of the continent; when they retreat into midlevel Canada, the weather will be unusually mild. In addition, a jet stream pattern that swoops down toward the southern United States in late summer can cause a flow pattern that can steer hurricanes my way; I have learned to keep a wary eye on its position. The jet stream boundary is the locus of squalls, cyclones, and storms.[17]

Jet streams can work as a steering mechanism for short-term winter weather too. In January 2005 warm, moist air from the southern Pacific flowed up the jet stream all the way to the Aleutians, then dumped massive quantities of rain on British Columbia, where they called it the "Pineapple Express." The same system followed the jet stream across the Rockies, where it became a clipper—Montana clipper in the United States, Alberta clipper in Canada. Snow and snow squalls duly followed. But that wasn't all: The system followed the jet stream into the southern Great Plains, then back up to the Ohio valley and out to sea somewhere around Virginia. After which, yes, it rode the jet stream back up past New England and Maritime Canada, bringing snow squalls and high winds. Thank you, Hawaii.

At high altitudes winds are organized into a sequence of high-pressure ridges and low-pressure troughs, with a wavelike motion. The largest of these wave patterns, the so-called standing waves, have three or four ridges and a corresponding number of troughs in a broad band. Shorter waves, several hundred miles in length, are called traveling waves. These form the upper parts of near-surface cyclones and anticyclones, and guide their movements and development. These high altitude winds are global in scope and are called geostrophic winds; they are driven by temperature and consequently pressure differentials, and are not influenced by the surface of the earth. These winds were unknown before weather balloons and are now generally measured by aircraft.

At low altitudes winds are essentially circular, organized into cyclones and anticyclones (low-pressure and high-pressure areas). As we have seen, cyclones rotate counterclockwise around lows in the northern hemisphere and clockwise in the southern hemisphere. Anticyclones revolve in the opposite direction. Extratropical cyclones are generally benign, little more than eddies in the overall system, but they are nonetheless significant: They are an essential part of the transference of excess heat received in tropical latitudes from the sun to polar regions;[18] without them the poles would be much colder and the tropics much hotter. In tropical latitudes the cyclones are smaller, usually not much more than 300 miles or so across, but they can carry winds of terrifying violence. When the sustained winds at their centers reach 74 miles an hour, they are called hurricanes or typhoons.

All in all, it seems like an orderly system, and it is. But on closer analysis it gets more complicated—much more complicated, for in other ways winds make up the most intricately beautiful and complex of the great engines that sustain life on the planet. Winds are steered and diverted and distorted by continents, mountains, forests, deserts, oceans, and large lakes, even cities, that muddy their flow and retard their passage. They also change winds' intensity. Land heats faster than water, and so localized pressure differentials are caused at every coast. Deserts, for their part, radiate heat faster than grasslands, and grasslands faster than forests, and each retains different degrees of moisture. All these factors complicate wind patterns.

Out to sea the patterns are simpler, and much more direct. If the world were perfectly flat—all ocean, perhaps—and didn't rotate, air would flow smoothly in perfectly predictable directions. Aerodynamicists call smoothly flowing air "laminar flow," as opposed to "turbulent flow"—we'd call it "streamlined." It is possible only where little interferes with the movement of air, and in nature this condition is very uncommon, so in practice pretty well any movement of air greater than 2 or 3 miles an hour is predominantly turbulent. It is one of the main tasks of car and aircraft designers to change turbulent flow into laminar flow to reduce resistance.

Large air masses don't mix easily. If they come together at speed, which is common enough, they set in motion complicated, swirling eddies of turbulent air. These eddies are what we call "weather."[19]

IV

Winds and weather are complicated in three other ways that make their understanding, and their prediction, much harder.

The first of these is the large-scale and relatively long-term cyclical fluctuations in the movement of air masses.

The second is the almost universal tendency of air to coalesce into vortexes. Hurricanes and tornadoes are the most obvious of these, perhaps because of their destructive potential, but there are many others, both useful and apparently useless.

The third is the purely local or regional winds affected by microscale topographic and geographic features that are not plottable on larger weather maps, but which nevertheless have significant consequences. Examples are the notorious *harmattan* of the Sahara, the *mistral* of the Mediterranean, and, closer to my home, *les Suêtes* of Cape Breton Island, whose astonishing accelerating effects on moving air I have experienced myself.

Take climate cycles. Perhaps the best known of these is El Niño, referred to by scientists as ENSO (for El Niño Southern Oscillation). Typical El Niño events, as they are called, last somewhere between six to eighteen months; their most obvious feature is a massive upwelling of warmer water in the eastern tropical Pacific, whose main consequences lie in the tropics but whose climatic fingerprints cause widespread droughts as far away as southern Africa and generally warmer winters in places like northeastern America. Its companion phenomenon is La Niña, which is its opposite: unusually cold Pacific temperatures. La Niñas occur after many, but not all, El Niños; their net effect is colder-than-usual northeastern American winters and warmer southwestern temperatures. As with dry years in the Sahel, there tend to be fewer hurricanes in El Niño years; the best guess is that one doesn't cause the other, but that some still-unknown factor causes both phenomena.

It's known that in the current global climate, El Niño years are warmer and La Niña years are cooler. It's also known that in 1976 the equatorial Pacific, potentially driven by anthropogenic warming, switched from a weak La Niña state to one in which El Niño occurs with greater frequency and intensity.

Does it therefore follow that more persistent El Niños would amplify global warming? Or that more global warming would result in more El Niños?

Perhaps, but not necessarily. A new study has found that in the early- to mid-Pliocene (5 to 2.7 million years ago), a steamy era that was the last time that global temperatures were warmer than they are today (atmospheric temperatures probably 10 degrees higher), the Pacific system was under an extended La Niña–like state, rather than the predicted El Niño one. These results were unexpected, and remain to be explained.[20]

El Niños were first acknowledged by fishermen from Chile, and because the phenomenon generally occurred around Christmas and brought them mostly beneficial results (more fish in the upwelling water), they gave it the name El Niño, which means *Christ child* in Spanish. La Niña, for her part, was originally referred to as *el Viejo* ("the old guy"), but was given its present name by the American media.

El Niños were first plotted by a British meteorologist, Gilbert Walker, in the 1920s, from as far away as India. Walker was trying to get a grip on what caused the often sizeable fluctuations in the strength of the Indian monsoons, and discovered that strong monsoons often correlated with severe droughts in Australia, Indonesia, and southern Africa. He also noted, without interpreting them, correlations between stable periods of high pressure in the eastern Pacific and periods of low pressure in the Asian Pacific. In his journals he called this "the southern oscillation." And there it rested, until the 1950s, when climatologists finally connected his hypothesis with what the Chilean fishermen had already observed. El Niño affects everything from large-scale climatic trends to microscale events, like wildflower blooms in the southern California deserts.

It is still impossible to predict when an El Niño will happen, a simple fact that makes climate-change skeptics raise their eyebrows— if you can't predict a simple recurring cycle a year or two in the future, how can you possibly predict climate change over hundreds and even thousands of years?

El Niño is not the only "oscillation" to affect winds and weather. There are at least a dozen others, and researchers seem to be discovering more every year. I spent a few months talking to atmospheric scientists and plowing through research papers in an attempt to understand their somewhat dizzying interconnections; at one point the wall in my office was slathered with charts labeled with impenetrable acronyms (AO, NAO, PDO, MJO, QBO, and others) of dubious utility, and eventually I tore them all down. Even to scientists, the impact of most of these cycles is only hazily understood.

Still, some things are clear. For example, the Arctic Oscillation (the "AO" in the list above) directly affects weather in the northeastern quadrant of North America and in western Europe, and tantalizing research has indicated some connections between the AO and tropical cyclone formation in the hurricane season. (There is also a subset of this cycle, called the North Atlantic Oscillation, but no one yet knows what it does.)

As its name implies, the AO circles the Arctic and extends high into the stratosphere. Its timescale is shorter than El Niño's—only a few months, or even weeks—and it cycles through a negative or cold phase, which brings high pressure to Arctic regions, along with lower-than-normal pressure over midlatitudes, and a positive or warm phase, whose effects are the opposite.

Perversely, warm AOs result in extracold weather in America and Europe. By contrast, when the AO high-level circulation is cool, it inhibits cold surface air dipping southward, warming up cities from Moscow to Vancouver, Calgary to Boston, London to Warsaw.[21] Similar oscillations exist in the southern hemisphere. Some studies suggest that the Antarctic Oscillation, the southern equivalent to the AO, was corrupted by the recent hole in the ozone layer, which resulted in extraordinarily cold winds, which in turn may explain why the southern polar regions were warming more slowly than northern ones before the ozone hole repaired itself. (Warming in Antarctica is once again accelerating.)

Two other cycles with implications for wind and weather are the Pacific Decadal Oscillation and the Quasi-Biennial Oscillation.

Climatologists rather like the PDO because it is a way of showing the general public that so-called normal climatic conditions can change, sometimes radically, over a period less than a human lifespan. The cycle often causes wild swings in Pacific marine species like salmon, and in local weather patterns.

It's possible the PDO is merely an El Niño writ large, with much longer cycles; there have been only two full PDO cycles in the last one hundred years—"cool" ones from 1890 to 1924, and again from 1947 to 1976, and "warm" ones from 1925 to 1946, and again from 1977 to the mid-1990s. It's also possible that the oscillation has two cycles, and not one—one from fifteen to twenty-five years, the other from five to seventy years. Its causes are unclear; the closest scientists can come is to suggest that it arises from some air-ocean interactions, which at least suggests some lines of inquiry.

Perhaps the most curious cycle of interest to wind scientists is the Quasi-Biennial Oscillation, in which the lower stratospheric winds of the tropics abruptly change direction, about every twenty-eight months. It is an enigmatic phenomenon, even by global oscillation standards. Why should winds that are easterly one month abruptly become westerly a few weeks later? And why should the easterly phase be about twice as strong as the westerly phase?

The QBO was only discovered in the 1950s, because it is only detectable at higher (stratospheric) altitudes. In the 1970s it was found that the periodic abrupt switches were caused by atmospheric waves starting in the tropical troposphere that travel upward into the stratosphere, where they are dissipated by cooling. The nature of these waves remains mysterious. The current culprit of choice is gravity waves, but what causes those and in this periodicity is opaque.

But they are important for millions of people, because hurricane activity is more common when these stratospheric winds are westerly, a pattern that is true also for cyclones in the Pacific. In its easterly phases, the PDO tends to knock hurricanes off their balance before they can really get going, while westerlies seem to act as catalysts. Why this should be so is another of the many unknowns.[22]

Reinforcing these oscillations of air, and probably partly caused by them, is a similar periodicity in global ocean movement that scientists call the thermohaline circulation, the steady movement of the world's oceans—with the Gulf Stream, the world's most powerful, and fastest, ocean current, as one of the prime engines. So important is the Gulf Stream that the whole phenomenon of thermohaline circulation is sometimes referred to as the North American conveyor belt. This conveyor belt has complex and still little-understood effects on winds and storms.

The conveyor belt is formed by water from the Florida Current, which circulates through the Gulf of Mexico and the Straits of Florida, and the North Equatorial Current, which flows westward along the equator. The resultant Gulf Stream parallels the coast of North America along a boundary separating the warm and more saline waters of the Sargasso Sea to the east from the colder, slightly fresher continental slope waters to the north and west. It more or less bounces off Cape Cod and is bent eastward, in the general direction of Ireland.

The Gulf Stream then feeds into the North Atlantic Current, which splits in northern and southern directions along the west coast of Ireland. The southward flow turns into the Canary Current, named for the Canary Islands off the coast of southwestern Morocco in North Africa, and thence bends westward parallel to the equator, winding up once more off Florida. Water flowing north along the west coast of Britain becomes the Norwegian Current, as it moves along the coast of Norway.

At least on the surface. Deeper down, the patterns are more complex.

When Gulf Stream water enters northern latitudes, it cools and sinks, becoming saltier and denser in the process (the *haline* in *thermohaline*). This happens in curious, slowly revolving "pipes" that take water from the surface to the seabed, mostly in the Labrador Sea and the Greenland Sea. At this low level, the water moves south and circulates around Antarctica; thence north again to the Indian, Pacific, and finally the Atlantic basins. The Smithsonian Institution estimates that it can take a thousand years for water from the North Atlantic to find its way into the North Pacific.

It is pretty obvious that changes in this massive circulatory device would have profound impacts, not just on wind but on climate generally. It is one of the main worries about global warming that increasing Arctic ice melt might alter or, worse, stop, the Gulf Stream, at least for a period. This indeed seems to be happening: The known vertical "pipes" have been reduced in number in the last few years from about a dozen to two, in part because the water is too warm to sink. The computer models all show that global warming would have a perverse short-term cooling effect on some northern places; instead of warming, Maritime Canada and northern New England, Ireland and the British Isles would go into a temporary deep freeze. I've sometimes contemplated this notion of my little house turning into Iceland, but it doesn't do to dwell too much on the possibility, because there are other, more immediately worrying things to be concerned about. For instance, a good deal of evidence suggests that changes in the velocity and direction of the conveyor belt might be a prime cause of the peculiar fact that hurricanes seem to wax and wane on a more or less thirty-year cycle. Is it just coincidence that the conveyor belt slowed down in the 1960s, cooling the North Atlantic slightly, in a period of fewer annual hurricanes? Or that the conveyor belt seems to have picked up speed starting in the 1990s, the decade when hurricanes started to increase in frequency again? Bob Sheets, former director of the National Hurricane Center in Miami, has asserted that the meteorological evidence suggests that the coming quarter century will produce more, and more intense, storms, and that "the thousands of people who moved to their dream homes during the hurricane low . . . could be in for some unpleasant time."[23] It's also possible that it might be the other way around: The frequency of hurricanes may, by contrast, affect the thermohaline circulatory system.

For us on the American northeast coast, therefore, this expected deep freeze would be a mixed blessing. More ice, but fewer hurricanes. If the Gulf Stream, on the other hand, moved farther north, we'd be warmer, but we'd be hit harder and more often by severe storms. When Hurricane Juan hit Nova Scotia in 2003, the ocean water temperatures were two or three degrees above normal. I asked Chris Fogarty, who has made a particular study of the relationship between surface water temperatures and hurricanes, what would happen if this increase persisted. "More Juans," he said laconically.[24]

Evidence is also accumulating that indicates the ocean's currents and hurricanes act together, in a feedback loop. The direction and speed of the conveyor belt can affect hurricane frequency, but the size and frequency of the storms can also push the Gulf Stream faster and farther. It was once thought that the prevailing winds alone, acting with our old friend the Coriolis force, caused the ocean's currents; it now seems that hurricanes have their part to play too. Kerry Emanuel, an atmospheric researcher at MIT, said in 2004 that the theory he had learned earlier in his career of how hurricanes began and sustained themselves was wrong. "It was always felt they were freak storms that didn't have much to do with climate in general . . . but we've come to different conclusions. Really, they are very integral to the climate system." His theory is that because hurricanes churn up the top 600 feet or so of the oceans, they lead ultimately to the circulation of all the world's oceans, which directly affects winds and storms, which in turn affect climate.[25]

This is the cheerful side of typhoons and hurricanes: Typhoons (from the Chinese *ta*, "great," and *feng*, "wind")[26] and hurricanes (from *Hunraken*, the Mayan storm god)[27] may be disruptive to a species that likes to build big-windowed homes by the sea, but within the complicated and coupled hydrographic-atmospheric system, they are nonetheless important agents of planetary self-government, redistributing air, moisture, and heat both vertically and latitudinally, scrubbing the air of accumulated pollutants, and accelerating the movement of the great ocean currents that keep our planet stable. A mature hurricane can export more than three and a half billion tons of air every hour, contributing greatly to redistribution of the troposphere, and can transport a billion tons of water over several degrees of latitude. They help make Earth work.

I should have thought of this when I was tracking Ivan. I mean to keep it in mind in the hurricane seasons to come. I just don't want a hurricane to drop any of that billion tons of water on *me*.

V

So much for climate cycles, the first of the three complicating factors in weather analysis. The second of these is the apparently irresistible need of flowing air (and liquids too) to form into vortexes.

By definition, a vortex is a rotation around a common center, often with a slow radial inflow or outflow superposed on the circular flow. As the air converges on the center, it begins to rotate faster and faster. "The process is surprisingly similar to an ice-skater's spin that accelerates as the skater's outstretched arms are drawn closer to the body."[28]

It's easiest to see in water—a whirlpool is a liquid vortex. Watch any stream that contains obstructions—a large rock, say—and you'll see a vortex, usually just upstream. I once watched a large log caught in just such an eddy, in a fast flowing river in British Columbia; it circled lazily in the eddy for more than three hours before its tip caught in the surrounding flow and it was dragged clear and went hurtling downstream (well, I had nothing better to do that day).

Vortexes occur throughout the universe, not just on our rotating little planet. Spiral galaxies are themselves versions of a vortex, caused by a combination of velocity and gravity. In our solar system, Jupiter's giant red spot, which is thought to be a long-lived vortex, is visible even to a small telescope. On Earth, cyclones are vortexes, and thus hurricanes and typhoons as well as tornadoes. Fire-caused whirlwinds are a hazard to firefighters. Dresden and Hamburg were both largely destroyed by fire vortexes. You can see vortexes in every campfire, in every little brook, in the water exiting a bathtub. Vortex motions, usually caused on Earth by pressure differentials in the atmosphere, are critical to a surprising number of human technologies, some economically important, others amusing but trivial. For example, airplane flight depends largely on vortex motion. Birds use tiny tornadoes to turn sharply in the air. The singing sound of high-tension wires is caused by vortexes. So is the ubiquitous slice of an amateur's golf ball, and, as we shall see when we get to Bernoulli's principle, even the annoying habit of shower curtains to billow inward and stick clammily to the person showering. A smoke ring is an example of a ring vortex. Some volcanoes blow smoke rings; so do steam locomotives. In nature, vortexes can have lifetimes ranging from a few seconds to several days.

In water, vortexes are called whirlpools, in which the flow is downward, and kolks, in which it is upward. Perhaps the most famous vortexes in history were Homer's Charybdis, off the Calabrian coast, and the Maelstrom, off Norway.

In air, the most common vortexes are whirlwinds and dust devils, which occur almost everywhere in the atmosphere, almost always accompanied by some degree of wind shear, or rapid interchange of air between layers; whole academic careers have been built on these boundary-layer studies. A typical whirlwind is the Australian cock-eyed bob, which picks up leaves, light twigs, and dust as it goes; the vortex winds we called *die duiwel gee om*, "the devil cares," near where I was born, were strong enough to carry tumbleweeds, some of them as large as hippopotamuses. Winds that build from the bottom up, like these whirlwinds, are sometimes called willy-willys; their more deadly cousins, tornadoes, are made from the top down, and of course are far more ferocious.

Tornado comes from the Spanish word for thunderstorm, *tronada*, which in turn comes from the Latin for turn, *tornare*, which is what vortexes do. And turn tornadoes do, spinning tightly with awesome speed. This is the most violent of all winds. Just how violent is still unknown, because tornadoes routinely destroy even the most robust of measuring devices, even supposing one could be placed in a storm's unpredictable path. But estimates have placed vortex winds at somewhere around 290 miles an hour, much faster than a Category 5 hurricane like Camille, and it is possible that occasional tornado winds might even exceed 480 miles an hour. The highest wind ever measured within a tornado was near Red Rock, Oklahoma, in April 1991, when a twister was clocked at 286 miles an hour.

The story is similar for barometric pressure. Standard barometers can't cope with the rapid changes in pressure caused by tornadoes, but pressure drops of 100 millibars are not uncommon, and drops of 200 not unknown. Because such drops occur in mere seconds, the normal pressure inside a building simply doesn't have time to adjust before the roof is blown off and walls are blown outward. The energy within a single tornado is not much less than the 20-kiloton bomb dropped on Hiroshima.[29]

I've had three near-encounters with tornadoes, and I've seen the results of others. The first was just a few years after I had nearly been blown out to sea in a Cape southeaster.

My family had moved to Johannesburg, a notorious locus of thunderstorms and massive hailstorms. One day we were in the family's aging sedan on our way from somewhere to somewhere else when the sky suddenly turned black, and then a violent yellow, and a deep rumble scraped across the city and across our nerves. My father brought the car to a stop and we saw the twisting funnel of a tornado touch down, perhaps a mile away, and then it was gone. Afterward he took me to see where it had been, a furrow of destruction two hundred yards wide carved through the edge of the city. Astonishingly, in the center of the path, a house still stood. Its roof was gone entirely, even the rafters, but nothing indoors was disturbed. Nearby was the remnant of a tree, only the first six feet still upright. The wind had driven a wooden clothespin deep into the wood so hard I couldn't pull it out. The family that lived in the house had taken refuge, as per the conventional wisdom, in the bathtub, and were unharmed. This being the South Africa of a certain period, there were servants quarters out back; these had been destroyed, and the woman who lived there had vanished and was presumed dead. The "savage and baleful Zephyrus," god of the winds of the sea, could not, surely, have been any more capricious than this.

The second encounter was in Arizona. I was at a conference of magazine editors, meeting in one of those gloomy, subterranean hotel conference rooms, when the lights abruptly went off. I remember it mostly because a publisher was whining about the ingratitude of her editor, and the rest of the assembled editors clearly felt the lights going out meant god was on their side, but then they came on again, and the discussion, such as it was, resumed. Outside . . . no more than five hundred yards from the hotel, a tornado had torn through. We were stunned at the proximity and the extent of the damage. The storm had ripped a path through town, leaving behind a ghastly jumble of mangled cars and demolished billboards, road signs, and small buildings. A pole with a traffic light still attached was poking out through the ruined windshield of a pickup truck. By some miracle, no one had been killed.

The third was in Ontario, which gets few tornadoes. At that time we owned some woodlands in the Ontario deciduous forest belt, and one day, when we were away in the city, a tornado tore through the woods not far from our cabin, so we saw it only by its results. Its path was peculiar, not uncommon with tornadoes. The whole thing seemed to have begun and ended on our small property. It tore through the forest for no more than a few hundred yards, but while it was there, it demolished a straight line of maples and beeches. It didn't just knock them over, as a hurricane would have. It tore them right out of the ground, roots and all, and pushed them into an untidy heap. No more than a few feet off its path, the trees were untouched. Even the leaves were still on their branches.

Tornadoes can happen anywhere, but the United States has the dubious honor of being far and away in first place, both in frequency and violence, with Australia an unenthusiastic runner-up. Other common-enough spots are the Ganges basin of Bangladesh and the Yangtze River valley of China; and I knew for myself that they happened on the great plains of South Africa, and occasionally in central Canada. Tornadoes, formerly just called whirlwinds, or occasionally, typhoons, were not unknown in Europe— Aristotle described a tornado in the handbook he called *Meteorologica*, and a tornado ruined parts of Rome in 1749. But the champ is America, where the tornado belt, or Tornado Alley, runs in a swath across the Great Plains from north Texas through Oklahoma, Kansas, and Iowa to southern Minnesota. Somewhere between six hundred and one thousand tornadoes touch down in the United States every year. May is generally the worst month. In its results, America's worst tornado was in March 1925, when a twister roared at 60 miles an hour through a series of small mining towns from eastern Missouri to western Indiana, killing 695 people. But in sheer perverse capriciousness, the unfortunate loser has got to be a small town in Kansas called Codell, which was hit by tornadoes three years in a row on exactly the same date, May 20, in 1916, 1917, and 1918. The tornado-free May 20, 1919, must have been rather a big day in town.

Squall lines, no matter how severe, seldom generate tornadoes, nor do normal thunderstorms—neither a squall nor a thunderstorm is a vortex, and to conceive a tornado the mother storm must show at least the beginning of a cyclone effect, a true vortex. The deadliest tornadoes are the creatures of mammoth and long-lived storms called supercells, whose winds are already rotating (they are themselves vortexes, albeit slow-moving ones) and may carry updrafts and downbursts exceeding hurricane strength. Some of these supercells can be 30 miles wide and 60,000 feet tall. Some of the other ingredients necessary for birthing tornadoes are warm, humid air near the ground, cold air at higher altitudes, and shearing winds. As with hurricanes, it is the humid air rising rapidly into colder air above that precipitates ice or rain, which in turn releases enormous latent energy, which then refuels the storm. Supercells almost always carry massive amounts of moisture, which often comes down as hail—many observers have reported what they call "hail roar" during a thunderstorm, the sound of billions of hailstones clattering together on the way to the ground.

Tornadoes can form very quickly, and are very hard to predict. Warm air rising rapidly into colder air above is a necessary precondition, but if the warm air rises steadily and smoothly, tornadoes are actually unlikely. A much more likely result would be another series of rather weak thunderstorms. But if a shallow layer of just-warm-enough air hovers above the surface—warm enough to prevent the ground-level air from rising—the potential is much greater for serious damage. Because if that cap is somehow moved or damaged, say by an incoming cold front, the pent-up warm air on the ground can burst through very rapidly. Then, watch out. Tornadoes can appear in less than an hour.

Fortunately, only one in a thousand thunderstorms becomes a supercell, and only one in about ten supercells causes tornadoes. The exact mechanism for tornado formation is obscure. They are more likely when surface winds blow in a direction other than high-altitude winds and the stronger the winds and the greater the height of the storm, the more intense the results. But as with hurricane beginnings, the actual tipping point is not understood. What is understood is that the Great Plains are the perfect kitchen for cooking tornadoes. This is because the eastern half of the continent is overlain in summer by warm moist air coming in from the Gulf of Mexico, and the western states, where the prevailing winds are westerlies, are very dry—they are in the rain shadow of the Sierras and the Rockies. Thus, Tornado Alley.

In the peak season, hundreds of tornado chasers (known, bizarrely, as "the chase community") spread out across Tornado Alley, usually in Kansas but also anywhere from Texas to South Dakota. Guessing tornado touchdowns is a sophisticated, if hazardous, game, and the Internet is full of more or less fanciful boasts from people who claim predictive powers that range from implausible to deranged. In high season tour buses packed with gawkers who want to experience nature's ferocity for themselves barrel their way down rural highways, hoping to get lucky. Some of these come closer and get luckier than they would have liked, and the occasional bus lurches out from under a supercell with windows shattered by flying debris or side panels dented by furious hailstones.

Some of these chasers are like the ham radio operators who bombard the National Hurricane Center with their track predictions, "useful fools," as they are often described by the professionals.

With tornadoes, because of their elusiveness and short duration, a curious symbiosis has developed between the chasers and professionals from places like Oklahoma's National Severe Storms Laboratory, who themselves fan out across Tornado Alley hoping to plant instrument packages directly into a twister's path. In practice, both "communities" keep in touch via cell phones and radio; they are occasionally plugged into emergency services and police bands when tornados are thought to be imminent. Tornado watches, which are little more than a guess at probabilities, are released to the public and the media several hours before tornadoes are expected, but warnings of actual twisters are released in a much shorter time frame. A network of Doppler radar units covers much of Tornado Alley, but the rupture in the cap that can produce tornadoes can be too small for the radar to easily see. As a consequence, sightings from the public, and from the tornado chasers, are taken seriously. With luck, warnings can come up to fifty minutes before the tornado strikes, but they can be issued as little as a dozen minutes before zero-point, perilously little time to take shelter, if, indeed, any shelter is to be found. Tornadoes seldom last more than an hour.

The first sight of a tornado is its funnel shape. It always seems to be striking downward at the earth, but this is an illusion. Tornadoes do form from the top down, but they aren't visible until they pick up debris from the ground—what you actually see is a grotesque mixture of earth, shrubs, fragments of trees, window glass, household effects, flying barns, bits of houses, whole cows, even car bodies and sheets of plywood and metal roofing. Their forward speed is usually around 30 or 36 miles an hour, but they can be nearly static or move well over 60 miles an hour. Their paths are usually narrow, no more than several hundred yards and sometimes less. Length varies widely from not very much to dozens of miles. The 1925 Missouri twister was huge, 9 miles wide and more than 180 miles long. A series of tornadoes crossed Grand Island, Nebraska, in June

1980 at a stately 4.8 miles an hour; the path of the final one included two complete 360-degree circles—the damn thing just
would not go away. Five people were killed. On April 22, 2004, a
tornado roared through Utica, Illinois, killing eight residents who
had taken shelter in the local tavern. "The sky turned purple and
then the air screamed," said one of the survivors, Mary Paulak. "It
sounded like a woman shrieking with rage."[30] The storm had struck
too quickly for residents to react. There had been fifty-one reports
of tornadoes the previous day, and everyone was nervous, but this
one came without any warning whatever.

As usual with tornadoes, there were anomalies in its 200-yard-
wide path of destruction. One house had its back walls sheared
right off, but a small cluster of cheerful orange tulips at the front
porch was untouched.[31] Anomalies, curiosities, quirks—these are
the nature of tornadoes. Almost every twister leaves behind some
curious fact—a children's doll driven feet-first into the trunk of a
tree, but otherwise undamaged; the whole roof of a house, still intact
with its gables and gutters, five hundred yards from the house it once
adorned; a car containing two children hurled into the air and back to
the ground, the children miraculously unhurt; a house demolished,
the one next door, a mere three feet away, untouched; blades of straw
embedded in fence posts; a schoolhouse with eighty-five pupils inside demolished and the children carried one hundred and fifty
yards, unharmed but seriously frightened; five railway coaches, each
weighing seventy tons, moved thirty yards . . .

A 1985 tornado in Barrie, Ontario, sheared a house in half, peeling it open like a doll's house. After it had passed, an ironing board
was still standing in an upstairs bedroom, the iron still on it, as
though ready for use. Of course I had seen for myself in Johannesburg how by cruel fate a twister had unfairly mirrored the
apartheid system by demolishing the quarters of a black servant,
leaving the main house and its white residents more or less intact,
and how a little wooden clothespin, made of soft wood and a short

length of twisted wire, had been driven hard into a tree. I'd also seen the aftereffects of a tornado that had passed through an oasis near Timbuktu in Mali in 1999, leaving the poorly constructed mud-built houses undamaged, but tearing out by the roots all the date palms, the reason for the community's existence. Leaving the houses alone was doubly unfair—they were of no further use. Less than a week after it had happened, the entire oasis was deserted. No repair crews were working away, no builders, no gardeners or planters, no herdsmen or householders. The oasis was empty, abandoned. The houses had been stripped, the camels moved off. Everyone had left.

Some of these anomalies are caused by smaller vortexes that spin around the edges of the larger one; videotapes of large tornadoes often show three or more smaller vortexes curling around the main funnel. Sometimes the strongest winds are generated in these associate twisters; wind tunnels have created small tornado-generation chambers that show how they work. In other cases, the quirks are caused by changes in the core circulation of the main body of rotating air—as it waxes or wanes, the ground-level effects can change from a few yards to tens of yards in seconds. Sometimes the funnel leaves the ground altogether, only to touch down again a hundred yards or so farther on—sparing one or two houses in a long row, with, always, apparently demonic unfairness.[32]

Less-violent wind vortexes than tornadoes are often just called landspouts. They're common at the higher elevations of Colorado and Kansas and in the Caucasus of eastern Russia, where the height of the land makes it difficult for strong tornadoes to form. Witnesses who have seen these elusive and fleeting apparitions describe them as curiously beautiful, almost luminous and translucent, perhaps because the low level of available moisture is not enough to fill them completely. They're called landspouts because they rather resemble their aqueous cousins called, for rather obvious reasons, waterspouts.

Water vortexes—waterspouts—are true tornadoes though their debris fields, and therefore their visibility, are rather different. Waterspouts have made their way into a good deal of fantastical literature—whole ships are said to be sucked up, and one early novel even had a stable community of spout-dwellers living comfortably at the spout's apex. Alas, even the more prosaic legends are mostly untrue. Waterspouts may pose some danger to small fishing boats, but none whatever to larger ships; nor do they suck up massive quantities of water, although they may lift water a few yards. They're really only visible because they contain clouds formed by condensation.

The most famous real, as opposed to fictional, waterspout appeared off Massachusetts in 1896. It was witnessed by thousands of holiday makers on a variety of beaches, for it appeared three times, lasting some thirty-five minutes. Estimates cobbled together from numerous excited eyewitnesses put it at over 3,000 feet high and maybe 250 feet at base.

It was widely believed that interrupting this column of air could be dangerous. "The violence of the wind retains the column in the air, and when that long spout of water comes to be cut by the masts or yards of the ship entering into it, when one cannot avoid the same, or the motion of the wind comes to be interrupted by rarifying the neighboring air with cannon or musket shot, the water being then no longer supported falls in prodigious quantities [upon the vessel]."[33] But no, contrary to legend, firing a cannonball into a waterspout will have no effect whatever, except to wet the cannonball on its way through.[34]

If you could interrupt a vortex, you would, indeed, destabilize it and cause it to fail; this is the theory behind controlling hurricanes. But the energy to do so is almost as great as that carried by the vortex itself, and the notion rather lacks for practicality.

VI

The third complicating factor in weather analysis is the microclimate winds, local systems that are geography and topography dependent, that ride on the back of global wind systems but that have a profound effect on local climate and weather. Engineers have to pay attention to local winds. The "local wind climate" can affect how buildings and bridges need to be designed.

But local winds have affected more than that. Winds affect not just myth and mood; they have also, in a very direct way, affected history, changing it for better or worse. If global winds have affected human history, in the sense of delineating those regions where cultures might best flourish, local winds and storms have affected it also, but much more abruptly—the "what if" school of history is full of weather-related stories. For example, a Saharan sandstorm foiled the Persian invasion of Egypt in the fourth century B.C. The Khanate assault on Japan was called off when a typhoon sank half the fleet in 1275. In 1529 heavy rains and high winds fatally delayed the progress of the huge Ottoman army under Suleiman the Magnificent, which would otherwise have captured Vienna and dethroned the Habsburgs centuries before their time. Sixty years later the Spanish Armada went down to defeat because the winds conspired with the British to blow in the wrong direction. When a gale blew the Armada back into port and one of his advisers suggested it was an omen from the Almighty, Philip II responded with what historian Geoffrey Parker called "naked spiritual blackmail": "If this were an unjust war," Philip declared, "one could indeed take this storm as a sign from our Lord to cease offending Him. But being as just as it is, one cannot believe that he will disband [the armada], but rather grant it more favor."[35]

A French fleet sent to sack Boston in 1746 was destroyed by savage gales, once off the Bay of Biscay, a second time near Halifax; the hapless commander, dispirited by the debacle, fell on his sword in his cabin, and retired from history. Many a time the fate of the American Revolution turned on wind. The 1775 "Independence Hurricane" drowned about 4,000 British sailors just south of Newfoundland, which in turn affected the British presence in New England in the months that followed. In 1776 George Washington called off what might have been a disastrous attempt to take Boston from Imperial troops because a "hurrycane" (in the word of one of his junior officers) intervened. The same year a violent Atlantic nor'easter allowed him to escape from New York. And in the following year he defeated Cornwallis in a crucial battle whose fortunes turned when a north wind froze muddy roads along the Delaware and enabled the Americans to reposition their artillery. Two violent storms sealed the fate of the British troops trapped at Yorktown in 1781. And more recently too: In the Second World War several campaigns were won or lost when the *harmattan*, the hot wind of the desert, knocked out communications. Not to mention the D-day invasion itself, which hinged for an awful moment on a break in an Atlantic gale. And the country of Bangladesh was birthed in a typhoon: In 1970 a tropical cyclone hit Bangladesh, then still called East Pakistan, killing more than 300,000 people; the central government's mishandling of the crisis that followed was a prime reason for the eastern province to break away and form its own state.

Like the global wind systems, local winds are a complex and shifting amalgam of factors, including the presence of bodies of water nearby, the accelerating effect of narrowing valleys, and the presence of mountains. These may be small, a particular valley in a prevailing wind, or very large, the Sahara Desert, for example, which is as huge as the continental United States. The most common local winds are sea and land breezes, because land and sea heat up at different rates, and mountain and valley winds (katabatic and anabatic winds)—the Foehn, the Chinook, the Santa Ana, and many others—because valleys and areas of rock heat up faster on one side or the other, depending on the sun's position, which results in thermal lift on the sunlit side and a cool downdraft on the other. These winds can be powerful. "Local" winds can also be quite large: The monsoons of Southeast Asia are really an overscale form of the sea and land breeze.

In a way, hurricanes and tornadoes, indeed all storms, are local winds too, although they are traveling winds with no fixed address.

In the Mediterranean, locally unique winds have been mapped for millennia. The fine-grained permutations are almost endless. Corsica, for example, has been known to record a Force 6 or 7 gale on the west coast, calm on the east, and quite a different gale in the Strait of Bonifacio, places only a dozen miles apart.[36] This has much to do with the presence of many islands, which distort and redirect winds. "The heat of an island creates new winds, mountains arrest some, accelerate others and all of these must meet, mingle and die or dance. The pairing of two islands means the re-diverting of already diverted winds."[37]

Wind anomalies exist, in short, everywhere. Within a few hundred miles of where I live are three of the most interesting. All three are due in some ways to the accelerating effect of local landscape, or what the forecasters, without apparent irony, call channeling. This is nothing more mysterious than the garden hose effect—if you pinch a hose, the water emerging through the pinch will come out faster than that going in. The same thing happens with winds. A small-scale example are the winds that are killing my rhododendrons. A somewhat more dramatic example is along the Presidential Range of mountains in New England, where Mount Washington for years held the record of the fastest wind ever recorded. The mechanism is simple: The westerlies race across Vermont, slide into the Connecticut River Valley, and then roar up the west slopes of the Presidential Range, compressed by the valleys into less and less space. At the summit of Mount Washington these winds exceed hurricane strength four days out of ten; in 1934 a weatherman recorded a wind speed of 231 miles an hour, faster than even the most terrifying hurricane. For decades this survived as the record speed for wind, until an observatory in Guam recorded a mountain summit wind of 236 miles an hour in a typhoon in 1997.[38] A wind-tunnel terrain study of Hawaii and Guam by NASA showed that wind funneled up a diminishing valley can eject peak velocities of about 250 percent of the flow coming in from the ocean.

The Canadian province of Newfoundland, a few hundred miles to my northeast, is one of the windiest places on the planet—it is surely destined, in the post-fossil-fuel era, to become to wind power what Saudi Arabia is to oil. E. Annie Proulx, in her novel *The Shipping News,* caught the unique flavors of the Newfoundland psyche when she described the wind scouring the bleak landscape:

> By midnight the wind was straight out of the west and he heard the moan leap to bellowing, a terrible wind out of the catalog of winds. A wind related to the Blue Norther, the frigid Blaast and the Landlash. A cousin to the Bull's-eye squall that started in a small cloud with a ruddy center, mother-in-law to the Vindsgnyr of the Norse sagas, the three-day Nor'easters of maritime New England. An uncle wind to the Alaskan Williwaw and Ireland's wild Doinion. Stepsister to the Koshava that assaults the Yugoslavian plains with Russian snow, the Steppenwind, and the violent Buran from the great open steppes of central Asia, the Crivetz, the frigid Viugas and Purgas of Siberia, and from the north of Russia the ferocious Myatel. A blood brother of the prairie Blizzard, the Canadian arctic screamer known simply as Northwind, and the Pittarak smoking down off Greenland's ice fields. This nameless wind scraping the Rock with an edge like steel.[39]

For most of its five hundred years of settlement, Newfoundland was populated only around the edges, in little villages called outports reachable only by boat. It wasn't until the coming of the railway in the late 1800s that people actually traveled over land. That had its drawbacks: Shortly after the Newfoundland Railway was completed, Passenger Train No. 1 was blown over during a winter storm and its mail car caught on fire.

The story is told by Mont Lingard, who wrote a number of books about the narrow-gauge "Newfie Bullet" trains, including one called *Next Stop: Wreckhouse.* Lingard, who had traveled the route many times himself, quoted the appalled engineer: "When we went down across Bennett's Siding [on the island's west coast, facing Labrador], we struck it. Just like running into a concrete wall. WHOMP! Mike had one of those big army parkas on. When he come out of the station, the wind picked him up and took him about twenty feet up in the air and slapped him down on the ground. He said, 'I nearly blew away.' The caboose and five cars went overboard. The caboose just cleared the bridge, except for that she would have went out in the water and we all drowned."

It happened again, a few decades later. As Lingard wrote: "Here was one of those big trailers, she was on the full length of the 40-foot flat car. She lifted up just the same as you lifted her up with a crane and moved her out about 15 feet clear of the track and brought her down. Never turned over or nothing, just lifted her up the same as you did it with a crane. It was unbelievable. They had tied her down and everything but that didn't matter. A good squall of wind took her and broke all that like nothing."[40]

The company that built the railways petitioned the government for money to build a new line elsewhere, but was refused. Since the company couldn't afford it, and could not afford to continue losing the occasional train to the nearby ditch, as was happening, it reached an ingenious solution: It hired a man called Lauchie McDougall as a human wind gauge.

Bruce Whiffen, who works for Environment Canada in its New-foundland office, has written an engaging account of the Lauchie story. The McDougall family had settled in the area in the 1870s. Lauchie was born in 1896 and lived his entire life at what was al-ready called, for fairly obvious reasons, Wreckhouse. He developed an intuitive and almost uncanny knack for figuring out the signs that indicated an approaching storm, especially the deadly south-easterlies. The railway company signed him on, paying him the then-substantial fee of twenty dollars a month, and over the years, Lauchie was credited with delaying hundreds of trains because of treacherous conditions. At least once, a train conductor spurned his advice and twenty-two cars blew over into the ditch. Lauchie died in 1965. You can still see a bronze plaque at the Marine Atlantic Terminal in Port aux Basques that says, in part: "This plaque is ded-icated to the memory of Lauchie McDougall (1896–1965). Mc-Dougall had extraordinary skills in determining wind velocities . . . through this area. Often called the human wind gauge, McDougall provided this service to the railway for over 30 years."

These appalling Wreckhouse winds were themselves caused by the garden hose effect. When an intense storm approaches New-foundland from the south, gusty winds converge along the south coast and extend inland several miles from shore. When they en-counter the Table Mountains a few miles north of Port aux Basques, the winds channel through the valleys and gulches between the mountains. The wind that then passes over and through the Table Mountains is far stronger than the original wind, sometimes devastatingly so.

The railroad no longer exists, but the Trans-Canada Highway was pushed through the same area, and in stormy weather truckers and car drivers are forced to delay their passage. Sometimes the more stubborn among them refuse; and the local police are resigned to pulling the shaken motorists from the ditch afterward.

The other local wind effect that I know well is even closer, no

more than a four-hour drive away, on the western side of Cape Breton, the northerly portion of the province of Nova Scotia. Cape
Breton is a lovely place, but its weather can be awful (it is a standing
joke around here that a forecast will say "windy today, with gales in
Cape Breton"). *Les Suêtes*, an Acadian corruption of the French
words *sud-est*, or south-east, are also channeled winds, but here the
effect is more complicated. *Les Suêtes* are so-called mountain waves,
which happen when stable air flows over mountains or hills and
combines with other effects such as drainage winds. In Cape Breton, southeasterly winds often blow ahead of an approaching
warm front. Above the surface and ahead of the front, the air is
very stable and causes a frontal inversion (warm air over cold air),
which pinches the approaching air stream between the inversion
and the surface. This has the by-now-familiar accelerating effect,
creating very strong surface winds along the coastal plains on the
lee side of the hills. If there are strong gale-force southeasterlies
inland and to the east everything collides, and hurricane force
gusts to 100 knots or more will be caused on the west coast and
ten to fifteen miles out to sea. In March 1993 an intense storm
starting in the Gulf of Mexico moved up the east coast to Maritime Canada, causing *Les Suêtes* winds measuring 126 knots, 150
miles an hour, in the little town of Grand Étang, lifting the roof
off Chéticamp hospital.[41]

VII

As meteorologists, and then mechanical engineers and the aircraft
industry, understood more and more about the turbulent and occasionally violent nature of air and wind, and as they came more and
more to see the importance of vortexes, a focus of these new wind
engineers became the physics of boundary layers, defined as the distance from a surface at which a velocity of wind reaches 99 percent
of the unobstructed "free stream." This might all seem too esoteric

for real-world applications, but boundary layer studies, combined with studies of turbulent flow, have generated a much greater understanding of how buildings and structures like bridges withstand the force of wind. It was the belated realization that wind's "energy content," in the scientific jargon, was affected by turbulence at the boundary layers that finally pushed wind engineering from something theoretical to something that would have a real impact on how buildings were designed.

The pioneers in this endeavor were Jack Cermak, a professor emeritus at Colorado State University in Fort Collins, Colorado, and Alan Davenport, who founded the Boundary Layer laboratory and wind tunnel at the University of Western Ontario, in London. Others include, most notably, Richard Kind, at Carleton University in Ottawa, whose research includes snow and sand movement in wind, wind-damage control mechanisms for roofers, and wind studies of retractable stadium roofs. But Cermak and Davenport are by far the best known. Cermak had built a massive wind tunnel in Colorado in the 1960s, big enough to model full-scale atmospheric boundary layers. In 1964 Davenport and his associate Les Robertson showed up at Cermak's lab to see if they could borrow his wind tunnel to check out designs for a new building project in Manhattan. They needed the best models they could get, because the project was difficult and expensive. They wanted to test everything—wind loads, pressures, and potential flexibility, everything they could. This new building would call for innovative structural concepts, and would be an incredible 1,368 feet tall. It was to be called the World Trade Center.[42]

The British-born Davenport has done the critical wind studies for many of the world's most complex engineering projects, among them the world's longest bridges and tallest buildings—not just the World Trade Center but also the Sears Tower in Chicago, the CN Tower in Toronto, the proposed 1,900-yard span crossing the Messina Straits in Italy, the Normandy bridge in France, the Storebælt bridge in Denmark, and the Tsing Ma Bridge in Hong Kong. The lab has re-created a miniature city and simulated wind conditions for Hong Kong, where turbulent typhoon winds put every building, including the Hong Kong Bank and the Bank of China, to the test. Davenport's résumé seems endless; he helped write the Caribbean Uniform Building Code, implemented in the 1980s, which is intended to render low-rise buildings more resistant to hurricanes. He is a founder of the insurance-industry-funded Institute for Catastrophic Loss Reduction, and the chairman of the steering committee of Project Storm Shelter, a facility sponsored by the International Association of Wind Engineering to improve the wind resistance of housing. And he is developing a series of high performance, low cost, prefabricated housing for use in postdisaster situations and other applications.[43]

It was a study of the proposed CN Tower in Toronto that cemented Davenport's reputation beyond his professional colleagues. The tower, a needlelike structure that dominates Toronto's skyline, was for years the world's tallest free-standing structure, a banal boast pretty much calculated to leave out guyed towers, of which there are several that are taller, and to finesse the fact that the tower isn't really a building but a communications tower with a restaurant on top. But since it was to be built in the middle of a city, and it would be by far the tallest concrete structure ever built, the architect's design was put to the test in Davenport's lab. It failed, and the original design was abandoned for a new-and-improved version. Davenport had shown that it would not stand up to the elements. "There were problems," Davenport says delicately, "with the amount of sway in public areas."[44]

Among the most important case-studies of local winds are those that involve long-span bridges. These are very complicated structures, susceptible to wind in many ways. They are liable to swaying, and to oscillations; the cables are liable to quiver dangerously, like massive violin strings, in high winds; and all components undergo stress and therefore fatigue. Oscillation is the enemy of bridges, and engineers must install what they call vortex-dampers to head it off. Out-of-control oscillations have destroyed a number of such structures over the years, including the 1836 collapse of the Brighton Pier in England, the 1879 collapse of the Tay Bridge in Scotland, the 1940 collapse of Tacoma Narrows Bridge in Seattle, and the 1986 Amarube Tekkyo rail bridge in Japan. Perhaps the most notorious of these was the Tacoma Narrows Bridge, which was captured on film—and which can be viewed on any of dozens of Web sites run by disaster junkies. The most interesting result of the collapse was to cause a new note of sobriety to enter the world of bridge design. Since the nineteenth century, bridge engineers had been besotted with new techniques and new designs, and suspension bridge designers competed with each other to achieve a maximum of structural grace and slenderness; artistic merit more or less overshadowed conservative engineering—a fact of life now endlessly drilled into the skulls of first-year engineering students. With its very shallow trusses and slender towers, the Tacoma Narrows Bridge was the high point in bridge artistry. Alas, it liked the winds far too much. It was shaped not unlike an aircraft wing—except that bridges are not supposed to generate lift. Only a few weeks after opening, the bridge had already developed the nickname Galloping Gertie for its tendency to heave in even moderate winds. A gusty windstorm in November 1940 was enough to do it in altogether. For an alarming few minutes, it twisted violently in the wind—at maximum twist one sidewalk was twenty-eight feet higher than the other—before a six hundred-foot section broke off entirely and plunged into Puget Sound.

All bridges are now subjected to full wind-tunnel tests before any concrete is poured or steel fabricated. Some parts of this testing are straightforward—the aerodynamics of the bridge cross-section and its towers are easy enough to model. Where it gets complicated is the introduction into the analysis of the "local wind climate," an incredibly complex study of historic meteorological records, historic wind directions and speeds, and the local topography—are there any accelerating topographical features around that would make normal winds into gales? And what about hurricanes? Not every area is hurricane-prone, but all areas might get hurricanes on rare occasions. How to model for those?

The problem with these local wind climate studies is that there is so much data, far too much to make accurate calculation possible: historic storm intensity data, storm track data, pressure differentials and the like, data on midspan and quarter-point pressures, defections and deflections, cable tensions, and many other variables, some of them with short-term periods and little apparent predictability. Even massive number-crunching computers are not up to the task; and even if they were, it would take far too long to input the data. And so engineers use a curious statistical sleight of hand with a whimsical name to provide answers about the sensitivity of the design to winds of various strengths and directions. They call it a Monte Carlo simulation.

As you might expect, the name is derived from the casino at Monte Carlo. Like scientists trying to model the real world on computers, gamblers too are faced with large sets of apparently random numbers. Each gambler, notoriously, has his or her own method of assessing the odds. In the casino as in the humdrum world beyond its walls, the numbers may be random, but very large sets of runs will provide statistical patterns that are more or less valid—as chaos theory would predict. To take the most widely known example, it is impossible to predict whether a coin toss will come up heads or tails, but a very large number of such tosses will always yield a fifty-fifty ratio of heads to tails. The Monte Carlo simulation, then, is simply a use of random numbers and probability statistics to investigate problems. It makes possible the examination of problems otherwise too complex for computation. For example, solving equations that describe the interactions between two atoms is fairly simple; solving the same equations for hundreds of thousands of atoms is impossible. Monte Carlo allows the sampling of such large systems, and the wind climate around bridges is a real-world example. (MC methods are used everywhere in science, in disciplines as diverse as economics and nuclear physics.)[45]

An intriguing test case for local wind studies was something not nearly as, well, *dire*, as the winds that destroyed trains in Newfoundland or the bridge near Tacoma. It was a wind that affected, and still affects, only the pocketbooks of a handful of very wealthy men. These are the winds around the eleventh green and twelfth tee of Augusta National Golf Course, the home of the Masters tournament in Augusta, Georgia. Golfers have ruefully called this Amen Corner, in recognition that only prayer seems to help the ball go where it is directed. Notoriously gusty winds from the left can be in the face of a golfer teeing off at the twelfth, but at the twelfth green, the target of the current shot, the flag shows wind coming from the right. How, then, to judge the shot? If you aim the ball to match the wind in your face, halfway through its flight it will abruptly be seized by a contrary wind and blown deep into the rough. If you compensate for that, you might still lose. The gustiness of the site means that the compensating wind late in the shot may not be there, dragging the ball into the rough on the other side. Tournaments have been won and lost at Amen Corner, which means substantial money is at stake.

Amen Corner is at the bottom of a slender valley between two hills. Tall trees surround the twelfth tee and green, in contrast with the relatively exposed eleventh green. For years, golfers have blamed those trees for causing the wildly eccentric wind directions at the three locations of prime interest: the twelfth tee and green and eleventh green.

In 2002 the Augusta National placed a call to Alan Davenport. His Boundary Layer wind tunnel researchers had never done a golf course before. It was hard to resist, so they took the commission.

The first trick was to construct a model of Amen Corner they could use in the lab. From topological maps, photographs, and sketches, they put together a 1:200 model made of high-density foam. Fairways were made of drywall compound; more than six hundred trees were constructed with sponge branches and wire-and-foam trunks; Rae's Creek was acrylic over a foam base, with silicone to replicate wind-induced wavelets. Tiny people were added for scale. Stage two was to model the surface wind speeds and directions. Data were available from a nearby airport dating back to 1949; from the raw data they extrapolated seasonal and annual frequency histograms corrected to the standard meteorological height of thirty-three feet. From this, wind speed and direction probability distributions were plotted for the full year and the month of April, when the Masters is played.

The next step was to plot the trajectory of a ball. They chose an eight iron for the twelfth tee. Data from golf ball manufacturer Maxfli showed that such a shot typically lasts just over 5.2 seconds, and its arc is known. The trajectory was shown by a thin copper wire coated with titanium-tetrachloride; a current fired through the wire created the necessary smoke.

Then various wind directions were simulated, together with the necessary gustiness, and the results plotted on digital video through a technique called stereoscopic particle imaging velocimetry. Because of the gustiness, which in reality is greater for the first few hundred feet of the ball's flight near the ground, no two visualizations were ever quite the same.

Still, the videos clearly illustrated the conflicting information anxious golfers must deal with and confirmed that the anecdotal evidence of wind behavior at Amen Corner is, in fact, largely true. The wind changed significantly along the shot trajectory. Near the twelfth tee, the wind is either directly in the golfer's face or slightly in the direction of the eleventh fairway. But near the peak of the trajectory the wind is moving more closely in the direction of the thirteenth fairway. At the twelfth green is a swirling flow with low wind speeds. If struck in the direction the golfer senses must be right, the ball will travel true for about a third of its path, about fifty yards. Then suddenly it would be hit by a strong crosswind from the left, carrying it depressingly out of plane into the thickets to the right of the fairway. For the last third of its flight these strong cross-winds diminish—they are still there, but weaker. Not one of these directions mirrored the prevailing winds of the day. As a consequence, each shot a golfer makes from the tee will be slightly different because of the natural gustiness. And that very gustiness made accurate prediction almost impossible.

The laboratory had provided detailed evidence that local topography influences wind patterns, and had shown how. A nice by-product was to provide architects, landscape designers, and farmers with their windbreaks and snow fences with more evidence that careful planting can mitigate wind damage and protect both buildings and crops.

The Augusta National, for its part, could solve the problem of Amen Corner by removing the trees, as they had suspected. But they had no intention of doing so. To watch the world's most skilled golfers turn occasionally into the rankest hackers was much more fun than getting out the chainsaws for a little silviculture.[46]

This was a classic case of "seeing" the wind through instrumentation and devising ways to deal with it. In just such ways, engineers can predict worst cases and best cases and use probability theory to devise protections against bad outcomes and uses for the good ones. Predicting when they'll happen, and at what intensities, however, turned out to be far more difficult. Variables can be subtle and hard to see; effects can be dramatic.

Weather forecasts are erratic for very good reasons.

CHAPTER FIVE

The Art of Prediction

*I*van's story: Ivan hurried into hurricane status
in three days, a full day faster than the official forecast. The transition came
early in the day of September 5, when Ivan was still 1,210 miles east-
southeast of the Lesser Antilles, at latitude 10° and longitude 45°. At five
A.M. eastern time maximum winds were already 75 miles an hour, just mar-
ginally a hurricane, but by the end of the day the pressure had deepened
sharply and sustained winds were already over 125 miles an hour—a strong
Category 3 hurricane, nearly a Category 4. Such rapid strengthening at such
low latitudes had not been observed before.

"Unprecedented" was the National Weather Service's measured word
for the phenomenon. It looked as though the people of the Antilles,
Puerto Rico, and Hispaniola all had a pretty good chance of a direct hit.
The official forecast, based on numerous computer models, skirted Puerto
Rico to the south and Jamaica to the north, and tracked right through
Haiti and the Dominican Republic. If so, it would miss Cuba's south
coast, but not by much. It was too early to say where it would make its
U.S. landfall. Much more than twenty-four hours out and the tracks became guesswork. Highly educated guesswork, mind you, but still guesswork. The "normal" recurvature to the northwest and then north—a very
common storm path—might not happen. A strong deep-layer anticyclone
(another name for a high-pressure center) was hanging about in the atmosphere to the north, which, if it remained in place, as was probable, would

*prevent Ivan turning at all, in which case . . . what? The center's fore-
casters consulted their models and reported in a National Hurricane
Center bulletin: "There is very good agreement through Day Five, with
NOGAPS to the right of the AIDS envelope and the UKMET to the
left of the AIDS envelope. The official forecast is slightly to the right of
the previous forecast track, and is in very good agreement with GUNS
and GUNA." I felt a bit like a kid with a new savings account whose
bank manager was blathering on about financial derivative products and
bond yields when all I wanted to know was how to take money out and
put money in.*

*I would find out about these models later and how they both
reassured forecasters and perversely, complicated their lives. But then the
best forecasters, as I would learn, were those people who managed to
shade guesswork into intuition, a very different order of information man-
agement.*

*Meanwhile, private yachts scattered into the Caribbean. A few went
south, perhaps forgetting the dire lesson of Hurricane Mitch in 1998, which
took a strange leftward turn at latitude 15° and slammed into the Honduran
coast, trapping dozens of small boats and the tall ship* Fantôme, *lost with
all hands.[1]*

*By Monday morning, September 6, Ivan was still plowing steadily west-
ward but had weakened some, to the middle ranges of Category 3, with sustained winds somewhere in the 120-mile-an-hour range. But the forecasters
weren't happy. They reported that Ivan had "improved his appearance"
and was now "better organized"—all of which anthropomorphizing meant
that while an earlier concentric eyewall had decayed, a newer eyewall, tighter
and clearer, had been formed. The storm was likely to strengthen, perhaps
back to a Category 4, but it would now likely pass to the south of Puerto
Rico and Hispaniola.*

*It was still being pushed southward by that protective subtropical ridge:
The storm couldn't get around it to head north. Not good news for Jamaica
or Western Cuba, but good for the Haitians.*

I

In December 1703 an extratropical cyclone of immense ferocity hammered England. It was the worst storm in English history, killing more than 8,000 people, toppling the newly built Eddystone Lighthouse and drowning everyone in it, peeling the roof off Westminster Abbey, and damaging scores of cathedrals and church spires, including the sublime Ely. Within the first six hours of the storm, which lasted almost a week, the Royal Navy lost twelve ships and 1,700 men, a fifth of the entire fleet; some 700 vessels moored in the Thames were driven ashore in a massive tangled heap.

In the storm's aftermath, the muckraking journalist Daniel Defoe, who had just been released from several years' imprisonment both for debt and seditious libel, placed a series of ads in the London papers soliciting first-hand accounts of the storm. He published the results in his best-seller called *The Storm*. He received dozens of eyewitness accounts of massive damage and a few of miraculous escapes. Queen Anne described the damage to the Royal Navy fleet as a "Calamity so Dreadful and Astonishing, that the like hath not been Seen or Felt, in the Memory of any Person Living in this Our Kingdom." The Reverend Joseph Ralton of Oxfordshire, for example, recounted seeing an immense "Spout, or Pillar, very like the trunk of an elephant only much bigger, crossing a field and, meeting with a great old Oak, snapped the body of it asunder, and coming to an old Barn, tumbled it down." Defoe himself counted 17,000 trees down before he grew weary of the tally, and 2,000 brick chimneys demolished. A part of Queen Anne's palace fell in with a great crash, and the "Lead, on the Tops of Churches and other Buildings, was in many Places roll'd up like a roll of Parchment, and blown in some places clear off from the Buildings; as at Westminster Abbey . . . and abundance of other places."[2]

From a wind point of view, the interesting thing about the great

English storm—a point made repeatedly by Defoe in his first chapter—was not so much the damage it caused, or what had happened, but that no one had even guessed it would happen. Put another way, everyone knew what had happened, but no one knew why.

It's not that people weren't interested, and weren't trying. Ancient weather lore was derived from careful observation of natural phenomena like clouds and by watching animal and insect behavior. The *Farmer's Almanac* notion that animals put on heavy coats before a particularly bad winter is not true; nor do squirrels increase their larders when severe winters are ahead; nor do frosts happen more often at full moon. But plenty of other signals are true, at least much of the time. For example, exposed seaweed will indeed swell ahead of bad weather, an effect that has to do with dropping atmospheric pressure.

Mariners particularly, whose safety depended on surviving storms, developed a whole litany of signals to foretell storms. Columbus, who in effect issued the world's first-ever hurricane prediction in 1502, based his anxiety on scudding wisps of cirrus clouds and on long swells from the southeast, signals of a storm that had nearly wrecked his expedition on his first voyage. Don Nicolás de Ovando, the new governor of the Spanish colony, ignored Columbus's warning, and ended up losing twenty-one ships to the hurricane that ensued.[3]

Columbus, like many a sailor before him and after him, recognized many different cloud types and shapes. It wasn't until 1803, though, that clouds were given their own taxonomy, when they were classified by Luke Howard. He arranged them into three general types: cirrus (curl), cumulus (heap), and stratus (layer). Other types of clouds, such as nimbus (rain clouds), were variations on these three basic types. Modern meteorology also classifies clouds by height—the prefix *cirro* denotes the highest clouds, 3.6 miles or higher; *alto* is the prefix given to clouds between 1.2 miles and 3.6

miles; clouds below that have no special prefix.[4] High clouds are usually made up of ice crystals, medium clouds of water droplets and ice crystals, and low clouds only rarely contain ice. Clouds, if you can learn to read them, can be accurate, if occasionally ambiguous, weather predictors. Light, scattered clouds alone in a clear sky usually mean strong winds. Clouds lowering and thickening always bring deteriorating weather, while clouds increasing in numbers, moving rapidly across the sky, are often a warning of really bad things to come. In the midlatitudes, scattered clouds with blue skies to the west mean predictably fair weather.

In short order sailors learned to watch for signs of a tropical cyclone. The first signal was a broad trail of cirrus clouds, with heavy showers of rain (it is now known that these travel 300 to 350 miles ahead of a storm). A sign that a storm was imminent was when that same cirrus departed rapidly, scattering in all directions, followed by a thin, watery veil that pervaded the air. By the time the next sign appeared—an ominous wall of cumulonimbus with layers going in different directions—it was too late to take evasive action, and the gear was lashed to the decks or stowed below, and stormsails hoisted.

This scattering cloud is a true signal, known to sailors everywhere. Sebastian Smith quotes a fisherman in the Mediterranean port of Carro, who told him the trick of predicting a mistral: "There may well be no indications of bad weather. The air becomes beautifully clear, but then you get clouds like little balls. Clouds like plates, or some say cigars, are another sign. But stay alert, when these balls start to explode and scatter, the mistral will soon be upon you."[5]

In the LaHave Islands and around the fishing town of Lunenburg, the same signals hold true. An overly clear day is ominous; when you can clearly see the branches of the black spruce on a headland across a large bay, be cautious. You can also smell weather coming:

If with your nose you smell the day
Stormy weather's on the way.

Experiments have found much truth to this old doggerel. High pressure that accompanies fair weather tends to keep scents dormant. When a low pressure system replaces the high, scents are released.[6]

Folk sayings have persisted for good reason.

Mackerel sky
Mackerel sky
Never long wet
and never long dry

Mackerel skies mean changeable weather.

If the moon's face is red, water ahead

This red can mean dust pushed ahead of the high winds of a low, bringing moisture.

Rainbow in the morning is the shepherd's warning.
Rainbow at night is the shepherd's delight.

A rainbow refracts light, breaking into colors; rainbows in the morning to the west usually indicate rain coming; at sunset a rainbow usually means the rain departing. As early as 1660 these portents were closely watched. "About noon we discovered one of those phenomena called a weather-gall or ox-eye because of its figure. They are looked upon commonly at sea as certain forerunners of a storm. It is a great round cloud opposite the sun and distant from him eighty or ninety degrees; and upon it the sun paints the colors of the rainbow, but very lively. They appear, perhaps, to have so great a lustre and brightness because the weather-gall is environed on all sides with thick and dark clouds." But this observation, by a Jesuit father crossing the equator, came with some skepticism: "However it be, I dare say I never found any thing falser than the prognostics of that apparition, I formerly saw one of them when I was near the continent of America, but which was followed, as this was, with fair and serene weather that lasted several days."[7]

> Red sky at night, sailor's delight;
> Red sky in the morning, sailor's warning.

Jesus, an early forecaster, made this prediction, or something very like it, in Matthew 16:2–3, to advise the fisherfolk on the Sea of Galilee, and sailors have been using it ever since. It sometimes works, and sometimes doesn't. Last night around my house there was a lurid red sunset across the bay, and this morning should have dawned calm enough for the lobstermen to be out setting their traps. Instead a warm front rolled through from the Midwest, bringing rain and blustery winds. A sailor named David Kasanof once expressed his skepticism that a sailor would show delight at any sign of redness in the sky, for redness denotes moisture. "Delight," he said, "is an inappropriate state of mind for anyone who has had the poor judgment to go to sea under sail. Under sail, I would suggest a state of alert suspicion bordering on paranoia as the appropriate mindset."[8]

In fact, in north and midlatitudes in the northern hemisphere, where weather systems move from west to east, red evening sky will bring clear weather about 70 percent of the time, and red mornings will bring foul weather about 60 percent. But in the Caribbean, where weather systems come from the east, the doggerel is useless.

Another old reliable in the northwestern Atlantic is the halo around the sun or the moon, a harbinger of bad weather. (Actually, a halo around the sun when bad weather is already here means it is over, and fine weather is coming.) These halos are caused by light refracting through the ice crystals in cirrostratus clouds; in a perfect example of apparently useless knowledge for its own sake, science has found that the crystals must all be hexagonal, less than 20.5 micrometers across, and producing a ray that is displaced 22°, no more no less. Larger halos, known as 46° halos, are produced when light enters through one side of such a crystal and exits near the bottom.[9] But the notion that halos are harbingers of bad weather is true, much of the time. Studies have shown that in two out of three times, rain or snow will occur within eighteen hours after a halo appears.

All these folk sayings, derived from long experience, are now much less useful than they once were. Atmospheric pollution has contaminated the signals and made them more erratic.

And of course satellite surveillance and accurate weather forecasting has made them, to some degree at least, redundant. A year or so ago I fell into conversation with a very old man on the wharf at Tancook Island, in Mahone Bay. His family had been lighthouse keepers in the area for three generations and he'd been a ferry boat skipper, and I figured him for a rich store of wind and weather lore.

"What'll the weather be this afternoon, Warren?" I asked.

"Oh," he said, squinting at the sky, "it'll blow a little, then some rain."

"How do you know?"

"Oh," said Warren again, as bland as could be. "I heard it on the radio this morning."

II

Weather forecasts are "right" somewhere between 50 and 80 per cent of the time. Depending on your definition of "right," and on the time frame, it could be worse than that, or better. In climatically unstable or variable regions, the five-day forecasts are notoriously unreliable, and it can be curious how often the fifth-day prediction seems to promise sunny or particularly pleasant weather. But before you sneer, think of the difficulties. The American Meteorological Society described it this way, its defensive tone entirely justified: "Imagine a rotating sphere that is 8,000 miles in diameter, with a bumpy surface, surrounded by a 25-mile-deep mixture of different gases whose concentrations vary both spatially and over time, and heated, along with its surrounding gases, by a nuclear reactor 93 million miles away. Imagine also that this sphere is revolving around the nuclear reactor and that some locations are heated more during parts of the revolution. And imagine that this mixture of gases continually receives inputs from the surface below, generally calmly but sometimes through violent and highly localized injections. Then, imagine that after watching the gaseous mixture, you are expected to predict its state at one location on the sphere one, two, or more days into the future. This is essentially the task encountered day by day by a weather forecaster."[10]

This more or less inspired weather-related guesswork goes back a long way into recorded time. It is known that the Mesopotamians, the people who gave the world the Hanging Gardens, were trying to correlate short-term weather changes with cloud cover and haloes around the sun and moon as early as 600 B.C. The Chinese, in their more formal and courtly way, tried to codify the weather even more, and around 300 B.C produced a calendar dividing the year into twenty-four segments, each associated with a particular weather pattern. Aristotle's four-volume *Meteorologica*, in which he dealt not just with wind but with thunder and lightning, hail and clouds as well, remained the standard text until the seventeenth century.

But weather forecasting as we know it, which is essentially a means of tracking wind and air systems and their effects, began in Europe, especially Germany, in the eighteenth century, when networks of towns shared weather observations. It started more formally in America and in Britain in the mid-nineteenth century. The Smithsonian Institution, newly founded, was by the mid-1850s having weather data telegraphed to its offices from those parts of the country that had been reached by railroad, and therefore the telegraph. In the same decade it cautiously, and very much after the fact, began to compile the first national weather maps. The Civil War put a brief stop to these efforts, but in 1865 a series of strong winter gales sank a number of ships on the Great Lakes, prompting the resumption of weather-data collection. A year or two later Cleveland Abbe, director of the Mitchell Astronomical Laboratory in Cincinnati, established a weather telegraphy service. The U.S. Army joined in, and then the Smithsonian again, and by 1870 President Ulysses Grant ordered the establishment of a formal army-run weather service.

Across the Atlantic, the British Meteorological Office was founded in 1854 as a small department in the Board of Trade, and by 1861 was already issuing gale warnings to shipping by telegraphing predictions to harbormasters, who would then hoist appropriately colored cones up a mast. Their forecasts were ostensibly for forty-eight hours, though they acknowledged that their day-two predictions were, to put it kindly, erratic, because most British weather came in from the Atlantic, where few observer stations were located. These forecasts persisted for a decade and then abruptly stopped, over the protests of the sailors who had been using their output.

The same thing was happening elsewhere in Europe, and by the 1870s data from weather observing stations all across the globe led to the construction of the first crude multinational weather maps. Which in turn led to the development of synoptic forecasting—the compilation and analysis of weather data from many different regions in the same period. In September 1874, the official weather map showed a hurricane for the first time.

Around the turn of the twentieth century efforts were made to develop what was called numerical forecasting—that is, forecasting the weather by solving mathematical equations that described the physical laws involved. This wildly optimistic notion—the real complexity of weather data had not yet been recognized—was first expressed by Norwegian weatherman Vilhelm Bjerknes in 1904, the year before Einstein wrote his paper expressing the special theory of relativity.

A short time later British mathematician Lewis Fry Richardson tried to put Bjerknes's ideas into practice and, working furiously for six months, produced a six-hour forecast for Munich. The futility of producing a forecast six months after the event happened was not lost on Richardson; nor was the fact that his forecast was wrong in almost all aspects. Rather bravely, Richardson reported on his failure in his 1922 book, *Weather Prediction by Numerical Process*. He suggested, tongue firmly in cheek, that the difficulties could easily be overcome: To predict the weather before it actually happens, you would need a roomful of people, each computing separate sections of the equations, and a system, not yet invented, for transmitting the results as needed from one part of the room to another. He guessed no more than 64,000 mathematicians would be needed.

The next real technical advance came in the 1920s, with the addition of huge amounts of high-altitude data. This was made possible by the invention of the radiosonde, a small lightweight box tricked out with weather instruments and a radio transmitter. Radiosondes were sent aloft tethered to helium balloons, climbing to almost eighteen miles before bursting. On the way down, they transmitted wind velocities, temperature, moisture, relative humidity, and pressure information to a ground station. Even now, for regular weather data collecting, radiosonde probes are the workhorses. Literally hundreds of them go up every day. Twice a day, every day, the little weather offices in places like Abidjan and Dakar and Niamey in West Africa, and on the Cape Verde Islands, and for that matter in Honduras, Cuba, a scattering of Caribbean islands, and yet again all up the eastern seaboard as far as Newfoundland, in Greenland and Iceland and the British Isles, helium- or hydrogen-filled balloons carrying a small payload of instrumentation are released into the atmosphere. At noon Greenwich time every day all this data is transmitted to regional offices, where there are any, then to national ones, and then the data flies across the oceans. Within minutes the computers in the national weather centers in Halifax and Miami and Ottawa and London and Hong Kong all have the same data to crunch.

The synoptic maps ("synoptic" here means a general overview) you see on your television weather channel or published in newspapers are based on this data.

Over the years, certain transborder conventions have been developed. Everyone measures the same things at the same altitudes. For example, data from all stations measures atmospheric pressure at 500 millibars. "Normal" weather would show this pressure at about 18,000 feet, so the critical datum is at what altitude and at what geographic node this pressure exists when the measurement is taken. If it is higher at 18,000 feet, this represents a high-pressure zone. If lower, it obviously means a low-pressure zone or front. The pressure gradients are shown on the surface maps by a series of curved lines called isobars. Where the lines are close together, the winds are very strong. Where they loosen up, the winds are light. On upper level charts, where the lines are closest together, are the jet streams. By the same consensus, weather maps show the wind direction as parallel to the isobars, with lower pressure to the left, looking downwind. Wind speed is calculated, in knots, by counting the number of isobars, analyzed at every 4 millibars, that fall within a spacing of 5° latitude, and multiplying by ten. For example, if 3½ isobars lie across 5°, the wind is $3.5 \times 10 = 35$ knots.[11] Consecutive sets of maps show the track of the different weather systems.

Numerical-calculator enthusiasts were given a new lease on life with the invention of the first computers in the 1940s. Late in the decade, mathematician John von Neumann at Princeton put together a team of colleagues and a few meteorologists to have another go at the problem. The team's director, Jule Charney, figured he could overcome Richardson's data swamp by using the computer and at the same time filtering out whole sets of data, such as sound and gravity waves. Indeed, in April 1950, Charney's group made a series of successful twenty-four-hour forecasts over North America, and by the mid-1950s, numerical forecasts were being made on a regular basis.

A decade later, on April 1, 1960, the first polar-orbiting data-collecting satellite, TIROS 1, was launched. It worked for less than four months, but it gave the world's weather people the first ever pictures of Earth and its cloud cover.[12]

Weather forecasting remained an art. But it now had real numbers, and real-time pictures, to back up its intuitions.

III

Part of the long struggle to understand and then to forecast weather was to accurately measure wind—and then to find some way of depicting it that others would readily understand. The new meteorological offices springing up across the industrialized world needed some way of describing wind force to their clients, at first merchant mariners and sailors, but then all kinds of industrial and shore-based users. As data collection became more widespread, standardization became more and more necessary. It was no use for a telegraph operator on the Prairies to refer to a gale, without some notion of what that gale meant, or how strong it really was. A gale may be a storm to some, just a fresh breeze to others. Notoriously, what fishermen from Gloucester, Massachusetts, considered a bracing wind would send yachtsmen from New York scurrying back to port—or that's what the Gloucestermen said, anyway. In any case, whoever they were, sailors needed something more precise than a "brisk breeze" (or, as some say of gales in New England waters, "a breeze o' wint") to describe what they were likely to experience.

By the late twentieth century scales had been devised for winds of all kinds—regular winds, hurricanes, and tornadoes. We even had scales for windchill and other esoteric matters. But the first of all these was the Beaufort scale.

This is one of these simple measures that seems to have been around forever. Most people know what it is meant to show—a simple numerical relationship to wind speeds, based on real observations of wind's effects. But its history is a little more complicated.

The eponymous Beaufort was Rear Admiral Sir Francis Beaufort, Knight Commander of the Bath, who was born in Ireland in 1774, son of a country parson, an émigré from France with a doctorate in law and a mischievous penchant for acquiring massive debts and then dodging the debt collectors. Young Francis apparently always expressed a yearning for the life at sea, and at the age of thirteen his father acquired him passage on an East Indiaman bound for China and the Indies by way of the Cape of Good Hope. That the vessel was wrecked in Java apparently didn't deter the young man, and he joined the Royal Navy as a midshipman on the *Aquilon*. His naval career wasn't exactly illustrious, but it was nevertheless quite successful, and by 1800 he had risen to the rank of master and commander. In 1805 he took command of the *Woolwich*, a forty-four-gun naval vessel that mostly acted as a supply ship for a force that was at the time attacking Argentina. It was there, in the chartroom of the *Woolwich*, that he devised the first version of the Beaufort scale.

"Hereafter," he wrote in his journal (now kept in a box in a filing cabinet in the Met Office in London), "I shall estimate the force of the wind according to the following scale, as nothing can convey a more uncertain idea of wind and weather than the old expressions of moderate and cloudy, etc. etc." The scale he devised ranged from 0, calm, to 13, storm. Within a few years he had allocated the numbers more specifically, to relate them in a way that sailors everywhere would understand. He had reduced the categories to twelve, which ranged from "light air, or just sufficient to give steerage way," through breezes and gales to "hurricane, or that which no canvas

Sir Francis Beaufort

could withstand." (For the full Beaufort scale and other wind measurements, see Appendix 2.)

So useful was this scale that by 1838 it was made mandatory for log entries in all Royal Navy vessels, and civilian vessels soon followed suit. One of the first recorded times the scale was put into daily practice was on Darwin's voyage on the *Beagle*.

Beaufort went on to become official hydrographer to the Admiralty, and died in 1857. His obituary made much of the charts he had devised for the Admiralty, and of his splendid work in ensuring safe passages for vessels everywhere, but made no mention whatever of the Beaufort scale. It was just one of his chores, something useful he did along the way. Useful, but at the time not worth recording.[13] It is still, however, being used, in one form or another. The curious thing about Beaufort's original scale is that it doesn't mention the wind speed. The only velocities mentioned are those that might ordinarily be achieved by a man-of-war under full sail at various conditions—Beaufort wanted his readers to look at the ship, not the wind. His numerical system was entirely arbitrary, but the ship's behavior each number was attached to was not—his descriptions would have been very well understood by sailors who had spent years in vessels similar to the *Woolwich*. They were all Royal Navy men who had put in their time blockading Europe and ranging to Africa and the Indies in sailing ships with quite recognizable characteristics. Beaufort's descriptions are couched in terms of the ship's behavior under sail—the first four for how easily a ship could be propelled, those of five through nine in terms of her sail-carrying capacity, and the rest for mere survival at sea.

The scale jumped from this—a shorthand reckoner for experienced sailors—to something still used by weather offices because of two small technological improvements and one accident.

The two gadgets were the first practical telegraph, invented by Samuel Morse in 1835, and the cup anemometer (wind gauge) invented by T. R. Robinson in 1846. The accident was more than just a small accident—it was a naval catastrophe.

In the fall of 1854 the French and English—for once on the same side—were fighting at Sebastopol when the fleets transporting almost all the supplies they needed for the coming winter siege were struck by a sudden savage gale. On the morning of November 14, the English lost no fewer than twenty-one supply ships, and the French almost as many, a maritime disaster that was rivaled in its intensity only by the almost complete destruction of a French invasion fleet aimed at New England more than a hundred years earlier, in 1746, off Sable Island, no more than two hundred miles from where I now live. The uproar that these losses caused was at least partly responsible for prompting the British Admiralty and the French Marine to jointly sponsor a weather network, in the hope of forecasting future storms before they could wreak similar havoc. This was the ancestor of the World Meteorological Organization.

Since the Royal Navy was involved, it made sense that their Beaufort numbers were incorporated into the new weather office data. But there were obvious problems—the new weathermen in Boston or Belfast or Bratislava, many of whom had never seen the sea, never mind a man-of-war, had trouble agreeing with each other on definitions. The result was confusion, made worse by the proliferation of wind scales. By 1900 there were more than thirty sets in vogue, some disagreeing by more than 100 percent. "It was no longer clear just what the old force scale meant, and few men survived who were competent to judge what the behavior of an 1805 man-of-war would be."[14]

The first solution was to produce a landlubber's version of Beaufort's observations, in which his "light air giving steerage way" was changed to "light air, direction shown by smoke but not by wind vanes," and his hurricane was changed from "that which no canvas can withstand" to the more stark "devastation occurs."

But in the end this wouldn't do either. In 1912 the International Commission for Weather Telegraphy began the search for real wind speed numbers to attach to the Beaufort observations. A set of equivalents was accepted internationally in 1926, and revised in 1946. By 1955, wind velocities in knots replaced Beaufort numbers on weather maps. At the same time, a gust was defined as "any wind speed of at least 16 knots that involves a change in wind velocity with a difference between peak and lull of at least 10 knots lasting for less than 20 seconds." A squall is more intense, "having a wind speed of at least 16 knots that is sustained at 22 knots or more for at least 2 minutes (in the U.S.) or one minute (everywhere else)."

And so we arrive at the modern Beaufort wind scale, which ranges from dead calm through light air (1 to 3 knots), to hurricane, measured at "greater than 64 knots" (74 miles an hour).

This was much more useful, and usable. Still, modern sailors often use Beaufort's own ship-oriented point of view. As Scott Huler put it in his book *Defining the Wind*, "Sailors tend to define the

Beaufort scale by simply looking at the sea—how high the waves are, what the surface looks like. As for the higher numbers—10, 11 or 12—who cares? It's just surviving anyway."[15] The fishermen on the eastern seaboard tend not to use a scale at all. They judge the wind by the swell and the break of the sea and by the sound of the wind in the rigging, and know from the pitch of the boat and the pitch of the sound when it is time to head for home. Even I have learned to judge a Force 9 gale ("41 to 47 knots") by the sound the spruces make in the wind.

If it is useful to know the speed, and therefore the effects, of gales and storms, how much more useful to have a measure of the planet's most powerful natural force, the Atlantic hurricane and Pacific typhoon? Curiously, though hurricanes were known in the Caribbean from the beginning of the exploration era, and though destructive hurricanes had battered the American coast for centuries (the 1900 hurricane that slammed into Texas all but destroyed Galveston), and though some of these storms actually made their way up the continent as far as the Great Lakes and occasionally back across the Atlantic to Europe, it wasn't until 1973 that a scale could be agreed on.

In the end, it was the construction industry and not the meteorologists that pressed for a solution, and the scale that was finally adopted was devised by Herbert Saffir, a building engineer, and Robert Simpson of the National Hurricane Center, which had been established in Miami. So the scale is called the Saffir-Simpson hurricane scale. It is a 1 to 5 rating based on the hurricane's intensity at the time of sampling. Category 1 hurricanes range from 74 to 95 miles an hour; Category 5s, the most severe, start at 155 miles.

It is a curious fact of wind that even in the very worst storms, and even in an open field with no obstructions, the velocity at ground level is effectively zero. When hurricane forecasters refer to surface wind speeds they really mean velocities at a standard height of 33 feet above the surface; from 33 feet wind is assumed to increase in speed with height, and generally does (see Appendix 7). The Saffir-Simpson scale, then, refers to sustained wind speed, measured over a full minute, at 33 feet. Gusts can be considerably higher.[16] Ratings assigned to hurricanes by the weather service are used to give an estimate of the potential property damage and flooding expected along the coast from a hurricane landfall, but wind speed is always the determining factor, as storm surges depend to a considerable degree on the slope of the continental shelf in the landfall region, on the elevation of the nearby land, and on topographical peculiarities—is there, for instance, the possibility of a funnel effect, which would push a surge higher than normal?

Anything above a Category 2 is considered a major hurricane, likely to do considerable damage to buildings and the landscape. (For the complete scale, and a representative sampling of major storms, see Appendices 3, 4, 5, and 6).

Only three Category 5 storms have ever hit continental America. The first was an unnamed storm that struck the Florida Keys in 1935, when the barometer fell to an unbelievable 892 millibars (26.35 inches). This Labor Day storm killed more than four hundred people; some of the victims were quite literally sandblasted, reduced to bones, leather belts, and shoes.[17] The second was Hurricane Camille of 1969, which struck the Mississippi coast with sustained winds of over 190 miles an hour and a storm surge of 25 feet above the mean tide levels—a three-story wave rolled through Pass Christian, knocking over apartment buildings, and one appalled survivor who had retreated to his attic was forced to break the window and swim to a nearby transmission tower, from which he saw the water submerge the peak of his roof. He lived two miles from the ocean. Camille is still the most intense storm ever known to have made landfall in America; the winds were so strong—probably 200 miles an hour from Long Beach to the ironically named Waveland—that entire sections of the Mississippi coast vanished. Although there had been plenty of warning, and evacuations had gone on apace, hundreds of people were killed and more than 14,000 homes completely destroyed. It probably didn't much encourage the survivors when President Nixon ordered the dropping of a hundred thousand pounds of the pesticide Mirex on the ravaged communities in an effort to destroy the plague of rats that followed.

Hurricane Andrew of 1992 was the third. Originally classified as a Category 4, Andrew blew away the antennae and radar disks of the National Hurricane Center, which subsequently, prudently, moved inland. The storm's rating was upgraded to a 5 more than a decade later, in late 2003.

Katrina, the storm that mauled New Orleans in 2005, was briefly a Category 5, but was Category 4 at landfall. Ivan, the storm I had been tracking, reached Category 5 not once, not twice, but three times. No other storm on record has done that.

A wind-intensity guide for tornadoes exists too. Because of their explosive but transitory nature, they are really only categorized after the fact, by the damage they have caused.

The scale was first written in 1971 by Theodore Fujita of the University of Chicago, together with Allen Pearson, then director of the National Severe Storm Forecast Center. It is called the Fujita scale, and for those of us with weather anxiety, it makes ominous reading, even if Fujita was, in the upper reaches, somewhat stretching the limits of adjectival vocabulary. Fujita 1, or "moderate" tornadoes, range from 74 to 112 miles an hour, and will peel off roofs, overturn mobile homes, and push cars off roads. Fujita 5s are described as "incredible" and their winds range from 261 to 318 miles an hour, "with strong frame houses lifted off their moorings, cars flying about, trees debarked and even steel reinforced concrete severely damaged." And yet Fujita described one grade above "incredible," which he called "inconceivable." These, if they ever occurred, would carry sustained winds of 319 to 379 miles an hour, but no one will ever know for sure, because all measuring devices would be destroyed, along with pretty well everything else in their path. (For a fuller description of tornadoes and the Fujita scale, see Appendices 9 and 10.)

A quarter of all tornadoes are marked as "significant" (F2), and only 1 percent are Fujita 3s or above, the most violent categories.[18]

Other useful measurement schemes devised in the twentieth century include the Mach number, the Reynolds number, and, most useful of all, at least in more northerly regions, the so-called wind chill scale. The Mach number, named after Ernst Mach (1838–1916), is used mostly by the military, by NASA, and, as a knowing aside, by former passengers of the transatlantic Concorde flights. It is, simply put, the speed of a moving object compared to the speed of sound; no winds, not even the most violent tornadoes, approach Mach 1. The Reynolds number, named after English engineer Osborne Reynolds, is on the face of it somewhat more esoteric. It's a way of expressing that how winds move through and around objects depends on their speed, density, temperature, viscosity, and compressibility. A wind's Reynolds number indicates whether the flow will be laminar (streamlined) or turbulent. It's widely used in aircraft and car design and to define wind flows around buildings.[19]

The wind chill, on the other hand, has immediate connections for anyone out in a winter gale. It has always been tricky to measure, and has generally been met with a certain degree of skepticism, because to some degree it measures how people "feel" in cold winds, rather than how cold they really are. But it is important, because winds do exaggerate coldness, and from about −35° Celsius, severe frostbite sets in within ten minutes—and much faster in high winds.

On calm days, the human body is somewhat insulated from ambient temperature by warming a thin layer of air close to the skin, the so-called boundary layer. Wind interrupts the boundary layer, exposing the skin directly to the air. It takes energy to warm up a new layer, and if successive iterations are blown away, the body feels— and gets—colder. Wind has another effect: It makes you feel colder by evaporating skin-surface moisture, a process that draws away yet more heat. When the skin is wet, it loses heat much faster than when it is dry.

The original wind chill formula was derived from experiments conducted in 1939 by Antarctic explorers Paul Siple and Charles Passel. Their table, first published in 1945, was devised by hanging a small plastic cylinder filled with water on a pole and measuring how long it took to freeze in various wind and temperature conditions. Their formula was modified somewhat over the years, but it still had two basic flaws as a practical guide. Human bodies shiver, for one thing—a plastic cup just doesn't react to cold in the same way. In addition, the measurement was done at the conventional height of 33 feet, where the winds are much stronger than at an average human height. So their table was unnecessarily severe.

In the 1970s the scale was modified by Robert Steadman of Texas Technological College in Lubbock, who proposed a scale that included not only wind but the intensity of sunlight, clothing worn, and other factors.[20]

And there it rested until the turn of the millennium. The push for a still more accurate measure, unsurprisingly, came from north of the U.S. border—Canada is not as dismally frigid as many Americans profess to believe, but, yes, it *is* colder. A Canadian-sponsored international symposium on wind chill attracted participants from thirty-five countries, and in 2001 a team of U.S. and Canadian scientists cobbled together a new wind chill index. Data were collected by the research agency of the Canadian Department of National Defence, which added to human knowledge by "volunteering" soldiers for a stint in a refrigerated wind tunnel, where they were exposed to a variety of temperatures and wind speeds. These hardy soldiers were, at least, dressed in winter clothing, but had their faces, the most vulnerable part of the human body to extreme cold, exposed directly to the air. The volunteers also walked on treadmills and were tested with both dry and wet faces.

The new index, now in use by both U.S. and Canadian weather offices, is expressed in temperaturelike units. The base comparison is to the way the skin would feel on a calm day. If the outside temperature is −10° Celsius, and the wind chill is listed at −20, it means that your face will feel as cold as it would on a calm day when the temperature is −20° Celsius. Or you can do it the other way: If the temperature is −10° Celsius and the wind speed is 30 kilometers an hour, the chart will tell you the wind chill is −20 (see Appendix 12). The wind chill, for the technically minded, is expressed in watts per square meter.[21]

Weather, wind, and storm projections first went truly public in Britain. The British Met Office began issuing forecasts to the public through the press in 1879. Its efforts were warmly greeted. Said the London *Standard*, "It may safely be conjectured that, unless the authorities of this most completely conducted department had already verified their forecasts within the not extravagant limits of time which are now mentioned, they would not assume this new responsibility."[22] This was somewhat overstating the case, as users were soon ruefully to understand. But the forecasts were useful enough that they have never been discontinued. And then, on January 11, 1954, the weather took to television, when George Cowling presented the first "in vision" weather forecast on BBC TV. He used an easel and "treatment to walls and background" that cost the Beeb £50.[23]

Well-coiffed weatherpeople have been with us ever since.

Almost every television newscast in almost every country now contains a synopsis of the current and predicted weather. Many countries now have channels specifically set aside for weather news.

Despite the cynicism with which these were initially greeted—after all, a weather channel demands drama to sustain its ratings, and drama means storms and probably exaggerated warnings about bigger, better, more frequent, and more violent storms at that—millions of people have come to rely utterly on their programming to plan travel and other activities. Even cynics, among whom I count myself, click their way to the weather channels when a major storm is thought to be imminent. In dozens of countries the national weather services uses commercial television and radio to disseminate their message; jointly, the weather industry has devised sophisticated but accessible computer graphics to mimic atmospheric weather conditions, and millions of people with only a modicum of scientific learning can now talk knowledgeably about isobars and frontal systems and follow the news about jet streams as eagerly as they follow whether Tiger Woods has won yet another golf tourney.

For those industries where weather is critical, commercial weather advisory services have sprung up. For example, a company called Compu-Weather in 1975 launched what it called "forensic weather services" in support of the insurance, legal, and engineering profes-sions. Its idea was to give careful analysis of what the weather had been at what was guardedly called "the client's point of loss," and it offered same-day service for busy litigators. It would also offer, for a fee, expert witnesses who could be expected to deflect property and liability claims. As an adjunct, it would also provide "24/7 weather decision assistance" for film and TV studios, "to deliver safe and ef-ficient on-location weather shoots." Will Tom Hanks need an um-brella today or not, or will his umbrella merely be blown away in a gale? Odysseus could have used a service like that.

IV

Because of hurricanes' enormous potential for destruction, predict-ing them has always been a special case of weather forecasting, and in many ways has driven the constant search for new and more so-phisticated weather technology. Available data were always depress-ingly sparse. Even after it was known that hurricanes were wandering cyclones, tracking their paths and intensities was at best a hit-and-miss business. At first, of course, this was because the proper tech-nology simply didn't exist. There were no high-flying aircraft or satellites, for one thing. Indeed, when the U.S. Weather Bureau was first set up there were no aircraft at all, and data radioed from ships was only incorporated into forecasts in 1909. But for a while the lack of proper data at the turn of the twentieth century, and the consequent failure to accurately track storms, was also a factor of the politics and personalities of the American weather service itself.

The best hurricane predictions at the time were those of the Cubans, led by the flamboyant genius Jesuit Father Benito Vines, who mixed intuition and meticulous observation to derive his often eerily accurate storm predictions. But although American relations with Cuba were then cordial, a turf war was under way in Washington—the head of the U.S. Weather Bureau, Willis Moore, had actually banned the use of the word *tornado* in forecasts, fearing it would cre-ate panic and through panic would come criticism, something he was ill-disposed to accept, engaged as he was in an attempt to centralize forecasting through his own office. That the Cubans often did things better than his people did was infuriating, and Moore instructed his people to ignore them or even to sabotage them. Vines died before the major storm that destroyed Galveston and killed thousands of its citi-zens, but his work was carried on by his successors, and the Cuban ri-valry with Washington persisted, with fatal consequences—the U.S. bureau discounted Cuba's alarming prognostications about the Galve-ston storm, and as a consequence failed to warn the Texans in time.[24]

Aircraft changed the way hurricanes were perceived. Radiosondes, the workhorse of the weather-prediction industry, were more or less useless in hurricanes. Hydrogen or helium balloons could only be released fore and aft of a hurricane—they'd simply be blown away otherwise—and so there was a yawning gap in the knowledge of what was actually happening inside major storms. Forecasters were restricted to land-based observations and the occasional report from a hapless vessel caught in the storm itself, though the crews were usually too busy saving themselves to spend much time updating the weather service. Radar, developed during the Second World War, helped to some degree. The National Hurricane Research Project was founded in the United States in 1952, and first used radar imagery to track a storm off Cape Hatteras in 1955. But radar was then land-based and fixed, useful for last-minute track changes but not for forecasting. Still, during Hurricane Isabel in 2003, the National Oceanic and Atmospheric Administration (NOAA) combined airborne sensors, offshore monitoring stations, and land-based radar to assess the storm. A series of towers with Doppler and SMART radars (Shared Mobile Atmospheric Research and Teaching) were deployed and were able to report minute-by-minute data back to the National Severe Storms Laboratory in Norman, Oklahoma—the eye wall actually passed between two of the towers.[25]

But it was really aircraft-deployed "dropsondes," first used in the early 1950s but not widely deployed until the 1990s, that revolutionized storm data collection. Dropsondes are aircraft-borne radiosondes; aircraft flying high overhead can drop them through a storm on small parachutes, and before they are destroyed, they can collect the same data the land-launched balloons do—pressure differentials, wind speeds, temperatures, and humidity.

The first storm photographed from space was Hurricane Ana, as early as 1961, but it wasn't until the late 1980s and early 1990s that satellite imagery from dedicated orbiters was employed to capture weather data, and for the first time forecasters could see, in real time, the actual patterns they were plotting on their maps.

More recently NOAA, the parent body of the weather service, together with NASA, launched a network of weather satellites with varying technologies and capabilities. One of the most promising of these, at least for wind measurement, is a technique called Synthetic Aperture Radar—SAR for short. SAR measures wind by calibrating every pixel value in the radar image to what is called "absolute radar backscatter." That is, it measures precisely the size and frequency of local image interference, and matches that to wind speeds and directions. A trial in the Gulf of Maine in 2000 yielded a finely grained map down to a twenty-five-yard resolution. The same year a satellite called RadarSat was launched carrying the new device, rather unimaginatively dubbed ScanSAR, which enabled scientists at the Jet Propulsion Laboratory to extract data for coastal winds over many hundreds of miles with a resolution of only hundreds of yards. Its use in hurricane-watching would be evident: It would enable the hurricane center to see at a glance small regions of very high winds, as a supplement to aircraft and other measurement.[26] Three new wide-swath ScanSARs were launched in 2002, the European Envisat, the Canadian RadarSat-2, and the Japanese ALOS.

Nevertheless, SAR is far from perfect. It has difficulty seeing through heavy cloud cover and in high winds, just the conditions in which accurate data become most necessary.

Another promising, though still in 2005 speculative, technique is the analysis of what are called "ocean microseisms." These are simpler than they sound: When seismometers capable of measuring ground vibrations were deployed decades ago, in the early part of the twentieth century, it became apparent that the ocean itself was giving off a continuous seismic hum, the product of the earth's response to wave-on-wave interactions. More recently it was realized that storms can be located and tracked using this seismic data. Because almost seventy years of archived information exists, "this approach allows, for example, the strengths of El Niño conditions to be assessed for times when [other] ocean data were unavailable."[27]

By 2004, space-borne "scatterometers" were slowly building up a real-time description of global wind patterns in a system calibrated by matching results with the evidence of wind buoys. A scatterometer is a device that sends microwave pulses to the earth, and uses the backscatter to measure the surface roughness. On land, the device has been used mostly to map things like vegetation cover in the Sahara-Sahel region, and to track shifts in polar ice. At sea, the backscattering is caused by ripples and waves, and can be measured down to a few inches. As the Jet Propulsion Laboratory (JPL) puts it, "the idea of remote sensing of ocean surface winds was based on the belief that these surface ripples are in equilibrium with the local wind stress"— which for nonspecialists means that the direction and height of the backscatter can tell you the direction and strength of the winds.

Scatterometers do not always work smoothly, either—ambiguities are encountered in interpreting wind direction, which require several casts at different angles to resolve, and rain can still fuzzy the images—but they are nevertheless the only instruments currently deployed able to give real measurements of ocean surface wind speed and direction under both clear and cloudy conditions, day and night. As W. Timothy Liu of the JPL wrote in *Backscatter*, "They give us not only a near-synoptic global view, but details not possible using numerical weather prediction models. Such coverage and resolution are crucial to understanding and predicting the changes of weather and climate."[28]

NASA first deployed the technology in a satellite called QuikScat, launched in 1999. The scatterometer on board used pencil-beam antennas in a conical scan, and was able to cover some one thousand miles in a continuous swath that reached 93 percent of the world's oceans in a single day. The device's standard resolution was fifteen miles, but repeat scans in special cases were able to reduce the resolution to about half that.

The satellite was put to early use. The National Hurricane Center had declared Hurricane Floyd to be a tropical depression on

September, 7, 1999, but QuikScat had found a surface vortex with the required wind speed a full two days earlier, and was able to track the vortex all the way back to the African coast—the future Floyd emerged from the Sahara to the sea on September 2. "Because such vortices, in their early stages, are too small to be resolved by numerical prediction models," Liu says, "and have no clear cloud signal, the scatterometer, with its high spatial resolution, is the best means, if not the only means, of early detection of hurricanes and the study of their genesis."[29]

There's a wonderful feeling of empowerment, and a somewhat more hubristic feeling of omnipotence, in looking at a satellite-eye scatterometer picture of the Atlantic Basin. In one taken after Floyd gathered strength and bulled his way into the Caribbean, and subsequently posted on the JPL Web site, you can clearly see the whole area dominated by a massive high-pressure system whose anticyclonic flow was creating strong northerlies along the coast of Spain and Morocco, implying strong upwelling in the ocean. Hurricane Floyd is clearly visible doing his business west of the Bahamas. And you can already see Tropical Depression Gert, later Hurricane Gert, forming a counterclockwise spiral in the central Atlantic. In the northwest Atlantic, off New England and the Canadian Maritime Provinces, nothing much was happening on that day. There would still have been tourists on the beaches around Cape Cod, though of course the satellite images didn't show them. Here and there, no doubt, lobsters were being boiled and broiled and roasted and consumed with the usual accompaniments, mostly beer mixed with beach sand. At that moment, our connections to the Caribbean, and to Mitch and Floyd and Ivan and his cohorts, seemed mercifully slender.

V

The difference between modern national hurricane centers is revealing. The American center in Miami is relatively new; located in an aesthetically challenged and charmless concrete and steel structure hunkered into the earth, its roof bristling with data collection devices, disks and domes and antennae of various description. So secure is the building that the only way to tell if a major storm is raging overhead when inside is to watch the computer screens that will tell you so—or to emerge like a startled rabbit into the open air above, to be buffeted about in person. It is a place built to withstand the most severe winds imaginable—and the denizens of the center can imagine and have experienced winds of awesome ferocity.

The Canadian Hurricane Centre in Dartmouth, which is governed by Environment Canada's weather service, is a very different beast. It was until recently at the top of a small eighteen-story highrise, for one thing. And "at the top" means just that—it is higher than the elevators go, and you have to trudge the last story up what looks for all the world like an emergency-escape stairway. While the U.S. center is like a bunker, the Dartmouth operational center is located in a large, airy room surrounded by curtain walls of glass, with magnificent views down along Halifax's extraordinary harbor past McNab's Island, which sits like a cork in its mouth, and thence out to sea. High-rises are not good places to be during hurricanes, and indeed when borderline Category 2 Hurricane Juan struck the city in 2003 the staff had to evacuate, not so much because they were alarmed for their safety but because the building's sprinkler system had failed, and thereafter bureaucratic regulations took over. Don Connolly, a Halifax broadcaster who was in constant phone touch with the center as he put the grim news out to his listeners, recalls blanching when he heard that the Hurricane Centre was evacuating. If they were bailing out, why not him? Through the large windows in his studio he could see the trees in the park across the street toppling in the gusts, but his own building was only four stories tall, and though it shook some, it stayed put. So did he.

The U.S. center in its bunker and the Canadian one in its eyrie reflect the realities that they each face. Miami doesn't get hammered every year but the chances that it will be are quite high. Dartmouth does suffer occasional hurricanes, and they can be more than usually unpredictable when they arrive. ("As hurricanes head north they become a different sort of beast," Chris Fogarty told me. "There is so much we still want to know.") Still, they are not common—the Maritime Provinces get a landfalling hurricane about every two or three years, and, if you include the region's maritime waters, one or two every year, but by the time they reach northern waters they are generally not much more than a moderate Category 1. Juan, which rode a northbound jet stream directly from east of Bermuda, scooting so quickly up the ocean that the colder ocean temperatures had no chance to spin it down before it hit land, was a small and extraordinarily violent storm, and the day after it blew off across the gulf of St. Lawrence rumors asserted confidently that it had actually reached Category 3, but in fact it just made it to a 2.

Despite their differences, the operating centers of both institutions are quite similar—computer workstations where the forecasters pass their shifts, color-coded maps showing the current season's activity matched to past history, and, everywhere, photographs of what hurricanes have wrought, a constant reminder that the predictions they are forced to make can affect not just the livelihoods but in some cases also the lives of the people they serve, and the usual jumble of papers, clipboards, and old coffee cups. The maps are color coded in the same palette the NHC uses for its public advisories—green for tropical depression, yellow for tropical storm, red for hurricane, each active track fronted by a bulbous nose spreading out from the storm's current position, indicating the zone of uncertainty in guessing its probable direction. These seventy-two-hour track and intensity forecasts are issued four times a day for all storms in the North Atlantic and northeastern Pacific. They show predicted longitude and latitude, intensity (maximum sustained winds), and predicted path to a tenth of a degree.

By November of 2004, the end of the hurricane season in the Atlantic, the year's map showed nine red tracks. Most of them, because
of a strong midseason ridge of high pressure, kept to a westerly
bearing and, graphically and ominously, converged on Florida. Ivan,
the storm I had been tracking, was shown clearly, but after it passed
through Florida, it lost its red color, even on the Canadian map.

On one wall of the Miami center, a composite has been pasted,
showing all the named storms from 1871 to 1998, a can of deadly
worms writhing across the ocean, menacing in its general predictability (most of the storms went west, curved north, and then
caught the prevailing southwesterlies) and even more so for the apparent randomness of each storm's path—some dived down into
South America, others headed out to the mid-Atlantic, the occasional one even bullied its way to Baja California; they hit Texas,
Louisiana, Mississippi, Florida, the Carolinas, Bermuda, New England, in a patternless wave of awesome destruction.

Sometimes the tracks even circle back on themselves, though this
happens more commonly in the Pacific—to the east of Australia,
where a regular procession of anticyclones arises every year, there
were nine single cyclonic loops, four double loops, and one triple loop
among ninety-three tracks, during a period of fourteen years. In the
Atlantic, Hurricane Jeanne did a strange double backflip in 2004
and circled around twice before abruptly haring out to sea.

A few African-born Caribbean systems even track back across
the Atlantic after catching the midlatitude westerlies. Some reach
Britain—London's 1703 storm, during which Queen Anne was es-
corted to a wine cellar under the Palace of St. James, and some
windmills were destroyed through friction-caused fire because they
were rotating so fast, is an example of what was probably an
African, then a Caribbean storm in its track, before it turned back
across the ocean. It had very likely completed its extratropical transition; it was no longer a true hurricane, but it would have coin-
cided with another low to cause a weather bomb—an explosive

pressure change defined as a drop of 24 millibars in twenty-four hours with a central pressure below 1,000 millibars, a state that can cause massively high winds very like hurricanes (bombs happen every year in New England and Atlantic Canada, usually more than once). Norwegian explorer Fridtjof Nansen experienced the remnants of a tropical storm in 1888 on the Greenland ice cap. The 1900 Galveston storm was still severe as it passed over Europe and disappeared into Siberia, where no records were kept of its passing. Some of these storms travel more than six thousand miles before they expire, causing havoc in an area several hundred miles wide.

Each workstation in each hurricane center is linked to the datastream coming in from quite literally hundreds of sources—weather buoys, remote sensing stations, aircraft and air traffic control weather data, reports from ships at sea, ham radio operators from Canada through Bermuda all the way across to the Cape Verde Islands, satellite and QuikScat data, Doppler and SAR radar outputs, and many others, including conclusions reached by hurricane and typhoon centers as diverse as those in Honolulu, Tokyo, Dartmouth, London, and Paris. There is even an acoustic model of a hurricane, containing data recorded from the apartment balcony of Dennis Jones, a scientist with the Canadian Department of Defence, and it's an eerie thing to listen to—you can hear the gusts breaking trees, for example, and you can "see" the gusts through their accelerated acoustic signature.[30] And each workstation has access through the broadest of bands to the various prediction models that have been put together so painstakingly for so many years.

These models, whose keepers are the Tropical Prediction Center and the National Center for Environmental Prediction, range from simple statistical tables to sophisticated three-dimensional equation simulations. They are of two kinds, track models and intensity models, designed to answer the two key questions: Where is the storm going? How strong will it be? These are the two most difficult—yet critical—things a forecaster must decide. Where is landfall to be? How rapidly will it intensify—or weaken? You know, because you have learned it to be true, that the maximum wind speeds are approximately proportional to the square root of the difference between the central pressure and the surrounding pressure, and the models will show you that. But now you need to know the speed of change, and no formula will yield that up. How to account for the explosive deepening that some storms undergo? Surface sea temperatures? The surrounding ridges and troughs? Something analogous to the intersection of wave trains that can cause rogue waves at sea? A jet stream crossing a storm's path at a critical angle? Before satellites and the models crunched on supercomputers, forecasters had relied entirely on data from ships, and to a lesser extent from planes, and matched those against known historic data; but their forward projections past forty-eight hours were frequently wrong by three hundred miles or more.[31]

Peter Bowyer, of the Hurricane Centre in Dartmouth, explains some of the difficulties: "A tropical storm is a disembodied unit in a much larger flow of air. Think of it as a cork in a river. The river meanders to and fro, and the cork will go pretty much wherever it is taken, wherever the river goes. In our latitudes"—he's talking of Atlantic Canada, and latitudes in the midforties to -fifties—"the flow of the river is very strong. In the tropics it can be very weak, sometimes so weak that it is no longer the dominant moving force. Then other factors, including rotation of the earth, come into play and can induce secondary rotations within a storm system, and those rotations can themselves become the storm's driving mechanism. So if the atmospheric movement above a hurricane is so weak it is hardly moving, just the difference in rotation between one side of the storm and the other will induce a forward motion and cause the storm to veer. Also, any kind of terrain, such as the mountains of Hispaniola, will change the course of the storm. A steering flow in the upper air is easier to predict because it is generally well defined."[32]

One of the critical factors in predicting Atlantic hurricane paths is the Bermuda high, the ridge of high pressure more or less permanently anchored in the midlatitudes. How strong it is, precisely where and how stable it is, will often be the critical factor in a storm's course. South of the ridge the prevailing winds are easterly, which is why storms off Africa head west toward America. They will tend to turn north because of the Coriolis force, but the Bermuda high makes it difficult to predict exactly where, or how sharply that recurvature will happen. Once through the ridge, storms hang a right into the southwesterlies, and then can move fairly smartly northeastward. Also, the westerlies can accelerate storms very quickly—they can travel from South Carolina to Newfoundland in twenty-four hours, which gives very little warning time to, well, people like me.

Here we have to risk getting lost in a thicket of acronyms: tracking models include CLIPER (from CLImatology and PERsistence), GFDL (Geophysical Fluid Dynamics Laboratory model), AVN (AViation Run model), NOGAPS (which, if you really need to know, is the U.S. Navy's "global spectral forecast model with 18 sigma levels, a triangular truncation of 159 waves, parameterizations of physical processes and a tropical cyclone bogussing scheme"; it uses complex motion equations to monitor air circulation around the globe) and UKMET (run by the British Meteorological Office, that is, UK MET). CLIPER is perhaps the simplest of the models, merely a historic track of hundreds of previous storms.

Models that attempt to predict a storm's changing intensity include the GFDL, which is also used for tracking storms, and SHIFOR (Statistical Hurricane Intensity Forecast), which uses climatological and persistence predictors to forecast intensity change. SHIPS (Statistical Hurricane Intensity Prediction Scheme) looks at the difference between the maximum possible intensity and the current intensity, the vertical shear of the horizontal wind, the persistence (that is, the previous twelve hour intensity changes), and other factors. A version of SHIPS is available for the Atlantic and East Pacific.[33]

Some of these models look only at radar data, others look to history, still others are based on broader, global, meteorological patterns. They all do it differently, and the forecasters, or weather analysts, will still have to make the judgment as to which model, or array of models, to follow in making their predictions. A forecaster has to get a feel for what works best in what circumstances, and not so well in others, and come up with a plausible synthesis. For example, most dropsondes measure the winds at the 10,000-foot level, and forecasters have to estimate how far to scale down estimations of surface winds—most multiply the figure by 0.9, but some use other measures, and the NHC has been criticized in the past for underestimating surface winds.

The forecasters are all scientists, with the data-processing training that the term implies, but they have also learned to rely on an ability to sense patterns from scattered information in a way that computers cannot hope to match. This apparent lack of rigor drives engineers crazy. How can you be creative and rigorous at the same time? The hiring profile for a weather analyst would be a mathematician, unflappable under pressure, fast to make judgments, well schooled in the models, and a good communicator. Analysts' judgments can have enormous consequences. When they draw the path-predictions on their computer screens, placing dots where they believe the storm will be in twelve, twenty-four, thirty-six, forty-eight, and seventy-two hours, it is this path that is issued by the center in its public advisories, and emergency preparedness services in a dozen places either stand down or go into heightened alert as a consequence; the same report might make the difference between thousands of people packing up to evacuate or staying home, a decision that can have dire consequences, sometimes life-or-death consequences. In the text explanations that accompany the bulletins, the forecasters can hedge somewhat and second-guess themselves. Many do, and will advise when the predictions are more than ordinarily uncertain, but they know as well as anyone that many of their readers only look at the path on the map and behave accordingly. "If I change one little thing in one of the models," Chris Fogarty told me, "it can change the landfall data by 100 kilometers [60 miles] or more. That can make the difference between destruction and escape." The forecasters have to learn to live with their guesses, to accept their mistakes, to coach their readers constantly not to take specific projections forty-eight hours or more into the future as gospel worth gambling on, and then . . . to do it all over again next time. Burnout, not surprisingly, is common.[34]

VI

Sometimes weather "customers," particularly those who are especially vulnerable to storm winds, need more hand-holding than national hurricane centers are able to provide. When you're out at sea in a small boat, you listen to the marine forecasts as often as you can. It can be wonderfully pleasurable sailing the open ocean in a small boat, but terrors lurk too, and yachts will scatter like chaff when they know a hurricane is coming. To supplement the official forecasts, a curious network of amateurs has emerged, people who provide direct and personal, and therefore doubly reassuring, links between weather forecasters and sailors. David Jones, a name nicely out of maritime legend, is one of them. He is a British accountant turned Caribbean weatherman who created the Caribbean Weather Network for yachters in 1993. Based on Tortola in the British Virgin Islands, he transmits on single sideband twice a day, at seven thirty A.M. and five thirty P.M. He gives the official forecasts for the

Caribbean but adds his own gloss, his own interpretation of a U.S. Navy forecasting model available on the Internet. In short order, yachting folk came to believe he was generally a day or two ahead of the National Hurricane Center.[35]

There are several other individuals doing the same kind of duty:

Schooner Arcadia *Schooner* Arcadia, *this is Southbound II Coastal. Do you copy?*

The voice is that of Herb Hilgenberg. He is hunched over a transmitter in the basement room that is the studio for Southbound II Coastal, his private commercial radio station. To his right, a computer screen is filling with an image of the globe as an updated satellite weather photo is downloaded, pixel by pixel. Over to his left, another computer is twittering away, compiling the raw data, isobar by isobar, that Herb has been transmuting into yachtsmen's gold—accurate weather data.

He toggles a switch. The airwaves hiss and crackle.

Schooner Arcadia, *do you copy, please?*

His face, amiable in repose, is furrowed in concentration.

"They're out there in the blackness of the ocean, all alone," he'd said that morning, speaking of his listeners. "The ocean can be very large, when you're all alone and an easterly gale is blowing. It's reassuring for them to hear a familiar voice." We'd been sitting in his sunlit kitchen, staring outdoors past the two satellite dishes tucked away in an L in the house, sucking data from the satellites orbiting overhead. "I'm talking to them directly; they know who I am."

Another sailor had said to me the day before, "Without Herb, you're blind. You don't know what's coming. It's like driving along a country highway without headlights. Herb is your beacon."

Southbound II this is Schooner Arcadia. *Good afternoon, Herb. How copy please?*

Loud and clear Arcadia. *What is your position and conditions?*

Our position is 37°24 N and 74°05 W, barometer 1012 and falling steadily, wind is northwest and light, wave height two feet from northwest.

The *Arcadia*, with captain Dennis Greenwood, is part of the great "they" that constitutes the devoted—not to say obsessive— listenership of the "Herb Show." "They" are the crews and owners of yachts and small boats from the Canaries to the Caribbean, from Venezuela to Newfoundland. Day after day, week after week, month after month, Herb sends his voice out into the Atlantic, and by do- ing so he helps to save lives, deal with and minimize crises, head off tragedies, track storms, and give early warning of dangerous lows and hurricanes. His listeners, the free spirits of the Atlantic, have come to trust him with their lives and their possessions. And though many Caribbean boaters have become wary of giving their exact position on air, for fear of hijackers and pirates, they know Herb needs impeccable information, and they give it to him. Many of them have learned to their cost that to doubt him is to risk having their boats mercilessly mauled by an Atlantic storm.

Arcadia, *I'm coming to you out of sequence.* Arcadia, *that second low I mentioned yesterday is moving into a position north of you. It is very in- tense and dangerous. I must urge you to alter course for Bermuda.*

Okay, Herb, thank you. We will head for Bermuda now.

On an average night Herb will talk directly to twenty, thirty, forty yachts. For every person he talks to, maybe a hundred are listening. Every night maybe a thousand people tune in to 12.359 (or some- times 8.284) megahertz to hear the steady, knowledgeable, compe- tent, informative, *comforting* voice of Herb Hilgenberg.

Sailor Hans Himmelman, from Halifax, put it this way: "When a sailor is down to his last amp of battery power, I'll bet you he's us- ing it to talk to Herb."

Herb is a weather router, and his single sideband radio broadcasts help ocean travelers navigate the always unpredictable weather of the Atlantic.

For this he gets paid—nothing.

And he does it from, of all places, Burlington, Ontario, a thou- sand miles from the sea.

Herb learned to sail in the challenging waters around St. John's, in Newfoundland, where he grew up, but he moved away from the Rock when he took his engineering degree and then an MBA from the University of Toronto. In 1982 he built his own boat and went to the Caribbean. "On the way, we got hit by a November storm. We were leaving Beaufort in North Carolina, and we weren't prepared . . . We had no idea it was coming. The U.S. weather service hadn't warned us. It took us six days to get out of it."

A few months later he got a job in Bermuda, "the center of the yachting world. It's the perfect refueling and provisioning port. And every week, you'd hear some horror story about a boat in trouble, demasted in a storm, its sails in shreds, taking on water . . . Caught unprepared."

Herb started to accumulate more and more knowledge about weather. "With an HF receiver, a single sideband radio, a modem, you can pick up a lot of raw data. I started to call out daily, but informally, forecasts just for my friends." In 1987 Hurricane Emily made a turn over the Turks and Caicos Islands. The weather services forecast it would stall and peter out offshore, but Herb disagreed. His data showed the eye would pass right over Bermuda. He battened down, told his friends to do the same. Finally, the U.S. Navy issued a warning at six A.M. The storm hit at eight. The eye passed just where Herb had forecast.

After that Herb became a constant presence in amateur radio circles and worked with the Bermuda Emergency Services Organization. Every Monday he'd be part of a routine broadcast about weather conditions, and, when hurricanes were about, he'd go on the air every six hours. People would listen. And so it went. "Two or three boats every day would talk to me. By 1989 I was running a daily schedule—I would cover people who became friends. In 1990 it became more hectic, and I was talking to fifteen or sixteen boats a day. It was taking time, more than two hours prep time and an hour on air. Even some fishing vessels in the Caribbean."

In 1992, Herb was talking to a couple of "his" boats in the general vicinity of the annual Newport to Bermuda yacht race. He warned them a gale was coming, though nothing adverse had been forecast. What he didn't know is that the U.S. Navy Training Squadron had a couple of boats in the race, and they took note of the fact that Herb, alone among forecasters, had got it right—and had undoubtedly prevented a few near disasters. After the race, the navy called and asked him to lunch. After that the NOAA people would download Herb's forecast daily—and Herb got access to sensitive satellite data he's still coy about explaining fully. It's no surprise, then, that hanging in his basement studio is a plaque from the Navy Training Squadron that says, "To Herb Hilgenberg, for Best Analysis of North Atlantic Weather and Sea Conditions."

That's where the "Herb Show" came from. In 1994 the Bermuda economy went sour and Herb found himself at age fifty-seven without a work permit or a job. The offers flowed in—come to Florida, to Norfolk, to Annapolis, to the Bahamas, Tortola . . . Give us your weather service, we'll look after you.

But Herb returned to his suburban Burlington bungalow, and within a few months was back on the air. Four hours prep time now, and three on air.

"It's even better here," he says. "The propagation is actually better. I can talk to boats all the way from the Cape of Good Hope, to Greenland, and over in the Pacific to Hawaii. I get e-mails, phone calls, from England, Europe. It's so *big*. At any time there are fifty, sixty boats out there, needing me."

For none of this does Herb charge money—that would spoil the special relationship he has with his clients, that of a kindly but, when necessary, scolding uncle. He does get voluntary contributions from grateful customers, enough to almost cover his $10,000 to $12,000 annual expenses, but he won't take money from people who expect a service in return. Those checks he sends back. Nor did he accept a contract from an astonished Lloyds insurance company, whose investigators had noticed that Herb's listeners tended to make fewer damage claims.

Here are a few excerpts from the *Arcadia*'s log:

Dec. 22. Afternoon. We turned for Bermuda as soon as we heard Herb's warning. We have learned not to ignore his forecasts. We arrived safely in Bermuda this morning and tied up inside Ordnance Island. A few hours later we were hit by the winds of one of the most intense lows the North Atlantic has seen for years.

Dec. 26. Afternoon. Steve and Mary arrived at noon. They left Buzzard's Bay two days ago. Their weather service hadn't warned them. Their boat was damaged and they were exhausted, safe only due to years of experience.

Dec. 27. Noon. At 5 am this morning the dockmaster awakened me to help dock a 65-foot sailing vessel. Its French crew looked haggard, exhausted, grim-faced. The main sail was shreds hanging in their rigging, the ship very battered. One of their crew was taken off by ambulance with a fractured neck. Hanging over the stern rail was the empty safety harness of a companion who had been lost overboard in a knockdown.

[Dennis, the skipper, added this unnecessary but heartfelt notation:] *This [man overboard] is the nightmare of every ship's captain. Once again I avoided it. Thanks to Herb. The French told me they'd taken their advice from a paid meteorological service in the States. They'd been told it was safe to sail.*

But even Herb can't win 'em all. One of his boats, the *SV Sparrow*, got caught in a hurricane. "We talked every twelve hours for fifteen days, it was exhausting. Finally the boat made it through. But in the end, the owner was so strung out that she fell asleep and ran aground on a reef, losing everything."[36]

Even when you know a hurricane is coming, even when its path has been accurately predicted and its wind speeds known, even then there are times when there is nothing you can do but hunker down and wait. The power of hurricanes is unmediated, unremitting, affected by neither hope nor belief nor artifice nor device.

Of course, humans being what we are, that hasn't stopped us from trying.

CHAPTER SIX

The Most Furious Gale

*I*van's story: On September 7, Hurricane Ivan
*briefly dropped down to a Category 2 storm, but as it pushed west past
Grenada in the Lesser Antilles it intensified suddenly and dramatically. The
central low pressure went down as far as 947 millibars and the winds in the
eyewall were estimated at 135 miles an hour, making it a Category 4.*

For the next two days there was a dissonance between the violent drama
of what was actually happening on the ground and the dispassionate analy-
ses of the forecasters' technical memos. In Grenada, twelve people died in the
storm; students at a local school spent a night wrapped in mattresses under
their beds as the roof and the windows peeled away with terrible shrieking
sounds; the seventeenth-century jail, picturesque from the outside but crum-
bling and overcrowded on the inside, was demolished and the prisoners, including the former deputy prime minister, incarcerated for killings in an
abortive 1983 coup, fled into the streets. The weather was so bad some of
them took refuge in a public shelter in Grand Anse, just outside the capital
of St. George's. Others took up machetes and went on a looting spree, perhaps intent on proving that their original sentences were just. Ninety percent
of the island's homes were damaged; the prime minister, who took refuge on
a British Navy frigate, had his own home completely flattened. Every major building in the capital suffered structural damage; even concrete structures became piles of rubble; wood and iron buildings often disappeared
completely.

The same day Ivan had slammed into Barbados and St. Vincent, cutting power and demolishing buildings; in Tobago, a pregnant woman was killed when a forty-foot palm tree smashed through her window and landed on her bed. The storm brushed by the Netherlands Antilles islands of Bonaire, Curaçao, and Aruba, flooded parts of the Venezuelan coast, and then headed off, at a leisurely 17 miles an hour, bearing northwest. It seemed to be heading for western Haiti. Or for Jamaica. Or maybe Cuba.

I was watching the technical discussions on the Web with intense interest. The meteorological context was complicated and difficult to predict. It was not just that it was coming hard on the heels of Hurricane Frances, which killed two people in the Bahamas and as many as fourteen in Florida and Georgia; the remnants of Frances were still causing flooding in the southeastern United States, those remnants separated from Ivan's track by that same stable ridge of a strong subtropical high, oriented southwest–northeast. The behavior of that ridge was crucial to understanding what would happen next. It could keep the storms separated. Or it could force Ivan into a radical turn to the right, steering it through Cuba and the Bahamas but thence harmlessly out to sea. Or, if the ridge lifted, Ivan could merge with Frances, with even more unpredictable consequences.

Still, the forecasters predicted possible passage over Jamaica and Cuba, and possible intensification. Any modification in intensity would come only through internal convective changes, entirely unpredictable, or through the tripping effect of a landmass. But the water ahead of Ivan was only forecast to get warmer—as warm as 30° Celsius south of Cuba and in the Florida Straits—and the storm needed warm water to keep going. After that, everything was the sheerest guesswork. But the track was unnervingly close to that of Charley, earlier in the season. Which meant Florida.

Again.

Early on September 9, a fine summer day with an early-morning shower and a warm sunny afternoon in the American northeast, Ivan drifted into the eastern Caribbean Sea. At five A.M. it was just north of western

*Venezuela, at 70° west and 14° north. The central pressure of 922 millibars
had dropped 15 millibars in seven hours, and a dropsonde into the eye
recorded surface winds at 140 knots. That made the winds 160 miles an hour,
and Ivan was upgraded to a Category 5.*

That was as high as the classification system went.

I

One way of getting a feel for the force of the wind, as I found to my
early sorrow, is to be seized by a gale and almost hurled into the ocean,
there (the childish imagination working in overdrive) to be dashed to
pieces on the rocks by waves the size of small mountains . . . But
there are easier ways too, and occasions much more pleasant. You
can stand, for example, on the pitching deck of a great sailing
schooner in a heavy wind, several acres of mainsail billowing out
above you, "taught and tight as the steel door of a safe," as actor
Sterling Hayden once put it, the wind caught in a capacious cup of
canvas to pull a hundred and more tons of vessel through the water
at better than 10 knots, bending the massive mainmast almost beyond its tolerances. I once read how in 1780 a British man-of-war
rounded the Horn in a gale strong enough to shred any vestige of
canvas, and, desperate to turn the vessel, the captain sent a dozen
men scuttling up the ratlines to the mainmast yardarms, there to act
as a small scrap of clinging living canvas, tiny in relation to the size
of the ship but enough, in that force of wind, to give the captain
some purchase against the storm.[1]

Sometimes, if your mood is right, such winds can seem playful.
When the wind is gusting to 80 and 90 miles an hour, scudding along
a beach, you can go out into the gale, spread your coat into the wind,
and bound down the sand in strides that are thirty and even forty feet
long, without effort. (Gerry Forbes, head of Environment Canada's
Sable Island Station, has done this many times as hurricanes have
passed by. "Of course," he says, "you then have to crawl back.")[2] At

120 miles an hour, the wind equals the earth's gravitational force, and then you will fly, whether you want to or not. Antarctic explorer Paul Doherty remembers how in a gale station staff would amuse themselves by going out into the hurricane-force winds, facing downwind, and then leaning backward, "leaning on the wind," at angles of 40 or 45 degrees. If someone then creeps up behind you and steals your wind, you'll fall flat on your back. "[A geologist] did this to me twice before I figured out what was going on," he writes. The first year he was at McMurdo Station on Ross Island, Doherty recalls, there was an iceberg one hundred miles long by twenty miles wide, "just about the size of the San Francisco peninsula." It stretched 1,500 feet down into the ocean, but the wind "pushed so hard on [the] . . . iceberg that it rotated around like the hour hand on a clock running backward."[3]

Great winds have their effects on the landscape too, not just on its inhabitants. I've seen the effects of aeolian erosion, which is what the geologists call sandblasting, in the Sahara Desert. Whole mountains have been stripped by the winds into grotesque castles, some of them with spires a thousand feet tall. Entire mountain chains heaved into being by tectonic shifting are reduced to dunes in the scouring winds. It's possible that the shape of the great pyramids of Egypt was inspired by natural erosion in the Western Desert. Geologist Farouk El-Baz has found that a pyramidal shape best resists erosion because it directs the wind smoothly upward, as does the prow of a boat, so natural pyramidal hills came to stand as a symbol of eternity. If the pyramids had been cubed, they would have disappeared millennia ago.[4]

Power and telephone companies in desert areas have to protect the lowest few feet of their wooden poles—the poles of the trans-Caspian phone lines lost half their diameter in a decade. On a tiny scale, I've seen a glass bottle carelessly discarded on the sand by a passing traveler turn opaque from the sandblasting in a day or two; in a month it too was dust and moved with the wind, dancing in little leaps, to disappear into the dunes. Wind is the most potent of all

erosional agents, trumping even water. It carries away somewhere between 200 million and a billion tons of topsoil and sand from the Sahara and Sahel every year, most of it dumped at sea but some
ending up in Europe or even North America. In the desiccated air
of the Sahara, wind-blown abrasive sand beats on metal with a
sound like heavy rain. It can rattle against legs and arms and face
like hail, causing abrasions and drawing blood.

Owen Watkins, a surgeon traveling with Kitchener's army toward Khartoum, once saw a train engulfed by such a sandstorm.
Maybe not quite as violent as the Wreckhouse winds of Newfoundland, but in his book *With Kitchener's Army*, he described it as
"quite the most grandly awful sight, a great bank of dust about 100
feet high, stretching for miles and miles in front of us and rushing
upon us with a dull roar . . . The [invalid convoy] train we had seen
start had to pull back into the station, and another train a few miles
out into the desert had to stop, being nearly swept off the line by
the force of the wind. After that experience I find it easy to believe
the stories, often read before in skeptical spirit, of whole caravans
being lost and buried in the desert."

One such story, recounted with commentary by Herodotus in
his *Persian Wars*, written about 450 B.C., told how Cambyses, the
conqueror of Egypt in 525 B.C., dispatched an army to quell the
stubbornly oppositionist Ammonites, keepers of the oracle at
Jupiter Ammon, at Siwa in the Qattara Depression in the Egyptian
Western Desert. "The men sent to attack the Ammonians, started
from Thebes, having guides with them, and may be clearly traced as
far as the [Kharga Oasis], seven days journey across the sand.
Thenceforth nothing is to be heard of them, except what the Ammonians report, that the Persians set forth from the Oasis across the
sand, and had reached about half way when, as they were in camp
breaking their fast, a strong and violent south wind arose, bringing
with it vast columns of swirling sand, which covered up the troops
and caused them to disappear. Thus, according to the Ammonians,

was the fate of this army." Forty thousand men, with their panoply
and pay chests, with their animals and their food stores, with their
armor and weaponry, with their commissary and their water skins,
perished in the sand, their skeletons polished and preserved, perhaps,
but never found.[5]

At sea the gales are terrifying even to experienced sailors; in the
stormy winter of 2002 to 2003, a container ship, a vessel fully seven
hundred feet long, limped into port at Halifax, wreckage still hanging over its gunwales; it had been caught in an Atlantic gale and had
fifty containers torn from its deck and dashed into the sea.

Winds heap up the water into massive waves. Oceanographers
say that it is possible—at least in theory—for waves of 220 feet to
be generated in the deep ocean, about the height of a twenty-story
building. That no one has ever reported such a wave doesn't mean
they haven't occasionally occurred; after all, hundreds of thousands
of ships have vanished since men began venturing onto the sea, five
thousand or so years ago. Slightly smaller but still gigantic waves,
however, are encountered too often for comfort. Hurricane Ivan itself churned up huge waves as it passed through the Gulf of Mexico, reaching as high as forty meters, or 131 feet, something that was
explained in a paper by Peter Bowyer and Allan MacAfee published
in June 2005 in the *Journal of the American Meteorological Society*. Ten
years earlier, in September 1995, the luxury liner *Queen Elizabeth 2*
was on route from Cherbourg to New York, and had to change
course to avoid Hurricane Luis. Nevertheless, the vessel encountered a series of seas 60 feet tall, with occasional taller crests. At four
in the morning—mercifully after even the most persistent revelers
had retired for the night—the Grand Lounge windows, 72 feet
above the water, were smashed by a breaking wave. Ten minutes
later, the bridge crew saw a wave dead ahead that looked, they reported later before the publicity-spooked owners told them to *zip
it*, as though they were heading straight for the white cliffs of
Dover. The wave seemed to take ages to arrive, but it was probably

less than a minute before it broke over the bow. A second wave, immediately behind the first, crashed over the foredeck, carrying away the forward whistle mast. The captain, R. W. Warwick, admitted in a report later that it can be difficult to gauge the height of a wave from a vessel, but declared that the crest was more or less level with the line of sight for those on the bridge, about 95 feet above the surface. The officers declared that it was not a swell but a true wave; and Canadian weather buoy number 44141 moored in the area recorded a maximum wave height of 98 feet at that time.[6]

High-latitude waves tend to be bigger and more ferocious than tropical ones, because cold air, which is denser and heavier, can raise higher seas at a set speed than warm air can.[7] Around Cape Horn, the winds lift the waves to awesome heights. Francis Drake's nephew, who was with his eminent uncle on one of his voyages, described "the seas, which by nature are of themselves heavy, and of a weightie substance [were] rowled up from the depths, even from the roots of the rocks . . . exceeding the tops of high and loftie mountains."[8]

Where ocean and shore intersect, so-called storm surges are caused in high winds. Much of the damage done by hurricanes is not through the wind itself but by surge-related flooding. I know this from my own property. When Hurricane Juan hit Halifax in 2003, it did massive damage to that relatively unprepared city. A mere hundred miles away at our house the winds were only a gale, but the ocean reared up like a beast, tearing at the shoreline. It thundered over our protective rocky beach and tore up our boardwalk, hurling it into the forest a full hundred yards away; the breakers were thirty-five feet tall, the storm surge almost eight feet, and the tide was high, another six feet or so . . . We get winter storms pretty well every year with hurricane-strength winds, but they seldom cause storm surges. One reason is that North Atlantic winter storms are bigger than most hurricanes—not in intensity, but in sheer spread. Their effects are therefore more diffused. The analogy is that a man with size-thirteen shoes leaves a shallower imprint in the ground than a woman with stiletto heels, which dig deep into soft earth. Snowshoes have the same spreading effect.[9]

The meteorologist's definition of a storm surge is "a complex deformation of the sea surface induced by the cyclonic winds on coastal waters, which surge as a sudden tide against the coast. The level of the sea can be raised by up to 10 feet for several hours. Depending on the characteristics and relative positions of the cyclone and the coast, the level of the sea can go up a further 3 feet by the low pressure."[10]

For many years scientists believed that the reduced atmospheric pressure alone could account for surges. But most of it is really wind-caused. A 35 millibar decline in pressure will raise the sea surface no more than a foot, and pressure-related surges seldom exceed three feet. By contrast, the storm that rolled through Galveston in 1900 raised the water level over fifteen feet; and Hurricane Camille raised the waters of the Gulf more than 25 feet.[11]

Even lakes can be affected. The Great Lakes of course, but one of the worst surges in memory was when Lake Okeechobee in Florida surged eighteen feet in 1926, drowning almost two thousand people. In the 1950s in Russia, Lake Baikal, the largest lake in the world in terms of volume, reared up in a storm surge and carried away an entire lumber mill.

On a greater scale still, as already noted, winds affect the oceans themselves. The great ocean currents, the earth's temperature regulators, are themselves caused by winds. Winds exert stress on the surface of the seas, and set entire oceans slowly spinning.

II

The critical thing to remember about wind is that the force it exerts doesn't just double when the wind speed doubles—if that were so, hurricanes would seldom cause any real damage, and on the other hand, windmills would only work in major gales. In fact, the inertial force of wind, the direct *push* of wind against an object, or the force exerted when something stops or changes the direction of moving air, is proportional to the wind speed squared.

A wind of 20 miles an hour will exert one pound of inertial force on a flat object 1 foot square. But this doesn't mean that the pressure will double if the wind doubles—that a 40-mile-an-hour wind will exert only two pounds of pressure instead of one. Instead, every doubling of the wind speed quadruples the inertial force, and a wind three times as fast exerts a force nine times as severe. You can see how rapidly this escalates if you take, say, the wall of a small shed some 400 feet square. It will face 400 pounds of pressure in a 20-mile-an-hour wind, 1600 pounds in a 40-mile-an-hour wind, 6,400 pounds in an 80-mile-an-hour wind and an astounding 14,400 pounds, more than 7 tons, in a 120-mile-an-hour hurricane. A building will have to be twice as strong to survive a 120-mile-an-hour wind as it would an 80-mile-an-hour wind. (See Appendix 11 for a wind force table.)

The effect is exaggerated by the so-called blast effect of a gust. In the case of a building, if a window or door suddenly breaks open in a gust, wind will explode into the building and destroy it from the inside. "No realistic amount of structural engineering can safeguard a building against the blast effect. It makes much more sense to take every possible precaution to ensure that windows and doors will not break or fly open during high winds."[12] But sealing the house tight against the wind may not be the right approach either. It may make more sense to enable some kind of controlled air flow through a building during a major storm. One homeowner whose house survived a major hurricane in Florida had closed his doors and windows tight and even sealed two turbine vents in the attic. The house was doing well until an errant tile smashed a window, and the wind immediately started to inflate the house like a balloon. He was saved only because the plugs in the turbine vents blew out before the walls did, and the pressure dropped at once.

Aerodynamics have shown that winds will exert an upward lift on any roof pitched less than 30 degrees, an outward force on walls parallel to the windstream, and a drag force on walls and gable slopes facing downwind. "In any major windstorm, then, some parts of a building will be pushed inward while other parts are being pulled outward. This combination of pushing in one place while pulling in another is particularly abusive and leads directly to many structural failures. Moreover, as wind direction shifts, and it usually does during any major storm, many of the forces reverse direction. Often a building will be weakened during the first half of a hurricane, then will collapse when the winds reverse their direction after the eye passes through."[13]

Interestingly, the pagoda style of roof design generates what wind-tunnel experts call a negative lift where they overhang, helping to keep the roof in place in typhoons. Which may explain why so many pagodas have survived for so long in typhoon-prone parts of the world.

A survey of Dade County, Florida, homeowners after Hurricane Andrew had passed by found that many of them had taken refuge in a bathroom or closet in the middle of a house. Such spaces, however, provided little more than emotional cover in a storm that was known to have driven two-by-fours through concrete walls. Much of Florida was considering adopting a code that would require new houses to have at least one room thought to be projectile-proof. Under consideration was an eight-foot-square room sheathed in two-by-four studs covered with four inches of plywood.[14]

Because hurricane winds increase sharply only small distances above the surface, what happens to high-rise buildings in high winds? Their steel skeletons can almost always stand the strain—though even the idea of being inside a high-rise that can sway more than twenty feet in a hurricane would make anyone queasy—but the winds could easily tear away the cladding and the curtain walls. The best-known instance was Hurricane Hugo in 1989, whose 135-mile-an-hour winds flattened thirty major downtown buildings in Charleston and ripped off the outside walls of waterfront condominiums. Tenants in a fifteenth-floor apartment said afterward they'd gotten out just as the outside walls went, sucking out all the furniture.[15]

Some of the most powerful winds, curiously, come directly downward—not just in tornadoes, but in so-called downbursts associated with thunderstorms. In fact, more damage is done by downbursts than any other kind of winds, perhaps because they are more frequent—a recent study found that more than two thirds of all high-intensity winds that did damage to buildings and structures came during thunderstorms.[16] Unlike hurricanes, the worst effects on downbursts are close to the ground, and so affect low-rise buildings more than highrises. Construction engineers and wind tunnel experts have been testing novel geometries and odd-looking protrusions on low-rises to reduce the wind-induced uplift of roofs. They include leading-edge spoilers, porous fences at building corners, similarly porous parapets, and roof-edge circular cylinders to induce downward flowing vortexes to counteract lift. In 2001, two researchers were granted a U.S. patent for "visually subtle spoilers . . . to achieve a reduction in peak negative pressures along the roof leading edge of low-rise buildings." Other research has shown that the best building to withstand high winds is a brick home with perimeter wall, roof, and balcony alignments all designed to provide the wind with a path of least resistance. Still, an experimental home built with precisely those factors in mind blew apart during Cyclone Tracy in Darwin, Australia, in 1974, along with half of the city. The cause—inadequate component fasteners. The building was a good idea; the attention to detail not good enough; the result complete failure.[17]

III

We now know what hurricanes and their typhoon cousins are. We know where they start, and some of the hows and whens, and all this knowledge is useful. We don't know much about the whys, though. Nor is it likely that we will ever be able to control hurricanes, though that hasn't stopped people from trying.

The first clues to the hows and whens of hurricane formation lie in the wheres—where they start and, more interestingly, where they don't. Hurricanes never start on the equator, for example—there is no Coriolis force at the equator, and so no way to get a storm spinning. They always start in latitudes just high enough for the Coriolis force to be appreciable. But they never start at high latitudes either—the ocean is too cool there, the sea surface temperatures, or SSTs, as the hurricane hunters call them, far too low. They may have their remote origins over land, especially where high heat and cool air produce thunder cells, but hurricanes proper never form on land—evaporated moisture is their fuel. Hurricanes hardly ever form in the southern Atlantic either—before they can properly organize they are broken up by the prevailing westerlies, which in the southern hemisphere are much closer to the equator, although in 2004, for the first time, Hurricane Catarina struck Brazil, the extraordinary product of high sea-surface temperatures, low vertical wind shear, and strong mid-to-low-level blocking currents. Few tropical cyclones start in the smaller Indian Ocean. Or if they do, they don't amount to much—the "fetch" of the ocean, the amount of sea available for a storm's nourishment as it travels, is too small.

Many of the Atlantic storms, like Ivan for example, are born in the Sahara, as the superheated air of the desert meets the cooler air over the mountains, and then is energized as it drifts out into the Atlantic—and so the weather offices in Timbuktu and Niamey and in Abidjan, in the Côte d'Ivoire and Dakar, in Senegal, are the early-warning systems for Atlantic hurricanes. Early-summer Caribbean hurricanes, though still made up of African-born tropical waves, tend to form in the western Atlantic and in Caribbean waters, because the sea there is shallower, and heats up faster. It is the mid-season hurricanes that pass over the Cape Verde Islands before heading their relentless way westward.

In the eastern Pacific, most hurricanes fizzle harmlessly in the colder water west of the Americas. Hawaii gets a few, but the prevailing easterlies steer most of them well south of the islands. The north and western Pacific is even more prone to cyclones than the Caribbean, and the "season" is yearlong and therefore much more dangerous than the six months or so of the Atlantic. Pacific typhoons are born at sea, and tend to be larger and better organized than their Atlantic cousins—the much greater stretch of the Pacific gives them substantially more room to mature than the smaller Atlantic. The classical Japanese description of a typhoon is *kamikaze*, or "divine wind"; Japan and the Philippines are frequently assaulted by three or more storms a year—Japan had ten in 2004. Korea, China, and Vietnam are also vulnerable. To the south, Australia's northern littoral is under threat from typhoons in summer and fall, from December through about May; about a dozen cyclones a year form offshore, and some of them strike land. The worst storm in Australian history was Tracy, which struck Darwin on Christmas Day 1974. New Zealand's waters are too cool for tropical cyclones, but like Maritime Canada the islands are prone to strong extratropical frontal storms as the cyclones wind down.[18] Pacific typhoons are tracked by storm centers in many places around the ocean's rim, but the most active are the Typhoon Center in Tokyo and a facility run by the U.S. Navy in Pearl Harbor, Hawaii. One of the oldest weather offices in the world was set up by the British in Hong Kong in 1884; the Chinese are still using it to track typhoons.

A precondition for the transformation of a line of thunder cells or a low-pressure wave into a tropical cyclone is stable air, a local environment where the winds don't change very much. Hurricanes don't form in patches of turbulent or active air—powerful as they are, destructive as they are, they are also curiously vulnerable at birth. There must be very little wind shear, allowing the big cumulus thunderstorms to build vertically. Any strong wind above a hurricane will destroy it, either by tipping it over through shear, or by literally poking holes in the warm core tube, allowing the warm air to vent, which weakens it. Or they'll "plug the chimney" at the top, where the storm's vent is.

Another precondition is warm water. The ocean below must be at least 26° Celsius, but preferably 26.5° or higher. No real scientific consensus exists on why this number is the magic one. It has perhaps to do with the climatological factors governing tropical oceans, but which factors and precisely how are still unknowns. Temperatures can be higher than 26°, but not lower—the higher they are, the greater the potential for damaging convection currents to occur. Higher temperatures don't increase the probability that a system will coalesce into a hurricane, but they do tend to make that hurricane more intense.

If these preconditions are met, the winds of the passing thunderstorms, still just a tropical disturbance or tropical wave, will evaporate this warm water, and because of the Coriolis force, the winds will lazily circle inward to the center. This causes a small vacuum, and the pressure drops, driving the warm, moist air upward. At about 2.4 miles above the surface, the vapor begins to condense into water, or into shards of ice and wet snow, and the act of condensation causes something deadly to happen: The heat, and therefore the latent energy contained in the moist air, is released. This energy is substantial—a kilogram of water will release enough energy to boil half a liter of water. This liberated energy then reheats the air, driving it still further upward, creating even lower pressure below, and drawing still more warm, moist air into the atmosphere. Consequently, the pressure drops still further, more heat rises into the sky, and the system, now beginning to spin faster, becomes a self-sustaining storm, isolated from the airstreams that surround it. If it remains coherent and well organized for at least twenty-four hours, that's when the hurricane centers of the world start to pay attention, because the precursor conditions for tropical cyclone formation have been met. If the sustained winds reach 23 miles an hour, the system is declared a tropical depression. Then it is given a number. Tropical depressions are tracked, because they are embryo hurricanes.

These tropical depressions are already convection engines, their fuel provided by the warm sea water, which evaporates ever faster in the higher winds, causing ever lower pressures. After a few days, the tropical depression may escalate into a tropical storm. If the sustained winds reach 38 miles an hour, the meteorologists reach for their naming dictionaries and give the new storm a moniker.

The practice of giving storms names rather than geographic locator numbers began in the final years of the nineteenth century in Australia, where forecaster Clement Wragge cheekily gave destructive typhoons the names of women he knew (or wanted to know), or politicians he thought were idiots. Bob Sheets, former director of the National Hurricane Center, credits a 1941 novel called *Storm*, by George R. Stewart, for bringing the practice to the Atlantic. Sheets notes: "It was easier, the hero said, to say 'Antonia,' rather than 'the low pressure center which was yesterday in latitude 155 E, longitude 42 N.' "[19] In World War II, military forecasters helped pilots to keep various systems separate with names rather than coordinates; the assigned names were random but always female, generally the names of wives, girlfriends, or, like Clement Wragge before them, women the forecasters hoped would be girlfriends. It wasn't until the 1970s that the practice was systematized. At that time forecaster Gil Clark drew up a list of women's names from baby-naming books and his own family. Men's names were added by the fiat of the World Meteorological Organization, under pressure from an indignant woman's movement, in 1978. Bob, a curiously boyish name for a major force of destruction, is the first male hurricane on record.[20] Gilbert, Clark's own given name, by coincidence still holds the record for the most intense Atlantic storm on record, with a low pressure recorded at 888 millibars. The world-record low pressure is 870 millibars measured in Guam in 1979.

At 74 miles an hour, when a tropical storm becomes a baby hurricane, pressure can drop very rapidly from the periphery to the center—pressure has been tracked to drop 38 millibars in thirty minutes in particularly severe storms. The energy is enormous—even a moderate hurricane releases enough energy in a single day to equal four hundred 20-megaton nuclear bombs; if converted to electricity, it would be enough to power all of New England for a decade, with enough left over to run toasters all over the Canadian Maritimes.

Meanwhile, the whole system is moving—for both Atlantic hurricanes and Pacific typhoons, generally westward at first, then curving northward and eventually northeast. This is why, in the northern hemisphere, winds are strongest to the storm's right, where the directional speed of the storm's travel is added to its rotational speed. To its left, the forward speed mitigates the observed wind speed. The faster the storm is going the more exaggerated this effect. If a hurricane is going to hit, you should hope you're to its left.

When a small hurricane's convection pattern strengthens, the centripetal wind flow gains speed, the input of warm moist air continues to escalate, even more large-scale condensation occurs during the ascent, and enormous amounts of latent energy are released, which in turn result in stronger winds, which in turn lead to uplift of more warm, moist air to condense and release ever more energy . . . A mature hurricane never blows itself out, as long as there is warm water to sustain it.

The evolution from depression to full-blown hurricane usually takes a good four days. It took Ivan not quite three.

It is curious that with all this data available, and with so much attention being paid to hurricane tracks and intensities, that the actual birth of a hurricane still remains invisible.

As I've said, we know where, we know when, and we know the necessary preconditions—but we still don't know why. What is the tipping point? What makes one system coalesce into a storm, another to dissipate? After all, a hundred Saharan thunderstorm systems drift into the Atlantic each year, but only a fraction even become tropical disturbances. Of those that do, only a fraction become storms, and not all of those become hurricanes. Globally, tropical cyclones are still uncommon. In any given year, the number will vary from thirty to one hundred, with somewhere around ten or twelve in the western Atlantic; of those, perhaps three or four will be defined as major. This may be changing: From 1951 to 2000, there was an average of ten named storms in the Atlantic each year; in the past few years the number has increased to about fifteen, and may still be going up—there are more tropical depressions to start with, and the ocean is two to three degrees warmer than in earlier decades. But the moment a storm becomes a hurricane is still hard to see.

No matter how closely scientists monitor the data, the exact instant eludes them. Only in what meteorologists call hindcasting, the after-the-fact scrutiny of the data, can they approximate it. The first hint is when they see wispy clouds lazily circling, drifting inward toward a point, an early sign of a system's struggle to overcome entropy, a sign of what they call, for obvious reasons, "organization"—a "well-organized storm" is a storm with serious potential. But even then, the why is mysterious. Storms are caused by dozens, perhaps hundreds, of forces that intersect and interact, sometimes directly, sometimes in ways so subtle they are hard to detect by even the most cunning of models. They have, in the scientific jargon, "a sensitive dependence on initial conditions." They are nonlinear, which mostly seems to mean they don't behave at all predictably.

In theory, it should be easy to track the beginning of a hurricane: Simply wind the film backward. We already know how its effects on the American coast (Step C) result from its westerly track across the Atlantic (Step B), which was caused by a tropical storm off the Sahara (Step A). Why not then follow it backward from C to B to A and then beyond to see how the whole thing began? With the vast array of data provided by the globe-encircling network of satellites, we seem to have plenty of information—those satellites can track phenomena to a resolution of a few yards. In practice, what you see when the film is unspooled is this: hurricane, smaller hurricane, tropical storm, tropical depression, thunderstorm, moist windy spot, then a set of weather conditions that in no way look any different from those that cause, well, nothing . . . What is it that energizes some of these warm moist spots into hurricanes? No one knows. All they can say for sure is that it must be very small, because it is presently beyond our ability to track.

Ernest Zebrowski Jr. in *Perils of a Restless Planet* explains how computer simulations of hurricanes, while crude, go some way to illustrating the phenomenon. A storm is created on screen using hypothetical initial data. Columns of numbers then yield wind speed, storm speed, barometric pressure, temperature, and other measurable variables. By itself, such a simulation yields no information of any value. But if you then conduct a second computer run, then a third and a fourth and a hundredth, giving each one a tiny variation in the initial data—say a few millionths of a degree difference in temperature, much too small to actually measure, or a tiny variation in wind speed—something surprising happens. For the first few hours of the simulated run, the "new storm" replicates exactly the course and intensity of the old one. But then, the behaviors of the two virtual storms begin to diverge, and eventually they differ quite radically—one might veer sharply north, the other continue on its westward course; one may die, the other bear down on Florida. Some theories suggest—though still unproven—that something as small as the effect of a flock of birds flying into the originating "warm moist spot" may end up changing a storm's history. Or perhaps the storm simply falls over in a light breeze. Or the low-pressure system may pass over a tiny atoll or island and be thoroughly changed in its nature. The errant butterfly of ecological legend may not be enough to change a storm's course, but if an embryo disturbance passes over a single resort hotel, that might well be enough.[21]

A scientist at MIT, Edward Lorenz, independently discovered the same phenomenon when he did the same storm-modeling run on two different computers, the old one using data to three decimal places, the new one to six. His model storm, too, produced major deviances after setting off similarly—a relatively tiny change, from 3.461 in the first run to 3.461154 in the second, produced large differences in both the storm's intensity and its predicted path. It was Lorenz who called the phenomenon "sensitive dependence on initial conditions," now more popularly known as chaos theory.[22]

Will we ever be able to achieve absolute accuracy in storm forecasting? Only, says Chris Fogarty of the Canadian Hurricane Centre in Dartmouth, if we have data inputs every few inches across the planet, both vertically and horizontally, something that is clearly impossible. With that many sensors, there'd be no room for people. And even if we were to blanket the earth with a sensor for every molecule, what then? The atmosphere contains more molecules than there would be electrons in any computer, so the calculations would end up being slower than the reality they were forecasting— you'd have a forecast that arrived after the event being predicted.[23]

The charts of historic hurricane tracks pinned to bulletin boards in hurricane centers everywhere need to be interpreted with great caution. Both the tracks and their numbers are truly unpredictable, classically chaotic. To speak of "increasing numbers of severe storms," or of "the storm of the century," or to predict the numbers that will occur in any one year is, in Zebrowski's phrase, to "entertain a statistical delusion." Because hurricanes, like other natural forces, are chaotic systems, you cannot predict how many will happen by looking at what happened in the past. There may be apparent patterns, but they are illusions. So if you know that from 1951 to 2000 the average annual number of named tropical storms was ten, six of which became hurricanes, or if you know that last year there were fifteen named storms and nine hurricanes, this average and this raw number are entirely useless to predict what will happen this year, or next.

IV

So great is the appetite for data about hurricanes that pilots have been flying into their deadly vortexes for more than sixty years, breaking through the maelstrom of the eyewall into the calm center of the eye itself, a beautiful and terrible place. They did this at first because they were told to (they were, at the time, military pilots in a wartime situation) and partly because the pilot's code of permanent insouciance meant that the challenge was irresistible. It was no surprise, then, that the pilots who did so referred to their flights as penetrations, and to the storms as females—the practice of assigning storms female names had started only a few years earlier.

The first deliberate penetration into an eye was in the summer of 1943. The pilot was Joseph Duckworth, at the time a flight instructor and a specialist in instrument flying, as opposed to visual flight rules. Part of his intention was to put to an extreme test his theories about flying correctly without, essentially, seeing anything outside the plane. Neither Duckworth nor his copilot, Ralph O'Hair, troubled themselves with keeping notes; apparently, undergoing the experience was justification enough. Still, they did confirm one of the existing theories about hurricanes, which helped confirm, in turn, notions of how storms persisted. They showed that temperatures inside a storm, in the eye, were twenty degrees warmer than outside it at the same altitude.[24]

The weather data these flights provided were useful, so useful that when the war ended, the military kept up its mission of weather reconnaissance flights. The most important squadrons were 57th Weather Recon based at Hicken Air Force Base in Hawaii, and the 53rd Weather Recon of the Air Force Reserve in Biloxi, Mississippi. The 53rd still exists, now part of the 403rd Wing, based at Kessler Air Force Base in Biloxi; they're called the Hurricane Hunters, and use ten Lockheed-Martin WC-130 aircraft. Other squadrons flew missions at Guam, Alaska, and Bermuda.

In the 1950s, Max (Maximillian C.) Kozak was chief warrant officer for the 57th. He was also a meteorologist, and after he left the Air Force found a home at the weather station at the Franklin Institute, in Philadelphia. His first time into a hurricane, as he recalls it, was a mission that ended at Johnson Island in the summer of 1955, through the eye of Hurricane Dot. He was the weatherman part of the crew of ten, stationed in the back of the B-29 bomber of wartime vintage, then the weather platform of choice, with his instrumentation and his single dropsonde. The briefing was to supplement whatever the ground stations already knew, to fill in the gaps in their knowledge of upper air storm data.

"Our mission was to fly daily," he recalls, "on general reconnaissance missions, mostly descending from 18,000 feet to identify surface craft in the area." Hurricane reporting was a sideline. Dot was their opportunity.

"Our crew was scheduled for takeoff at midnight. Dot was just off the big island of Hawaii. Light rain was falling as we took off and climbed to 18,280 feet. The idea was to 'box' the storm, collecting data from its periphery outside the 50-mile-an-hour wind band, and to penetrate the storm—meaning penetrate the wall cloud into the eye—at sunrise. The first go-around was no problem, but on the second go-around the navigator spotted something on the radar. He called me over. I got out of my seat and saw a saberlike line of activity moving with the winds northward. I knew what it was.

"I said to the pilot, 'Can we go around this?'

"But the tail of the anomaly extended eastward, around 50 miles or so, so the pilot said, 'No way. There are rocks in those clouds.' We'd be over land. The rocks were the mountains on Big Island.

"But I knew what we were heading for, so I said, brace up, we're going to hit a down draft. Everyone strapped in, and we just waited. For ten minutes we flew horizontally, normally. Then, sure enough, down we dropped, suddenly, a thousand, three thousand feet in just seconds . . . I thought for sure the wings had been pulled off and looked out the window to see . . . and there they still were. As I looked, I saw a sort of undulation upward. Somehow, we got a boost of wind from below and climbed out."

Max recalls that the second navigator, a hurricane virgin, didn't take the drop so well. "He began to shout and yell, and as we called it 'fell out of the tree.' We had to subdue him, so we tied him up and put him under the navigation table where he fell asleep."

The flight continued to box the storm as it moved steadily northwestward. No one made notes. They couldn't. The turbulence was too severe. No more heart-stopping thousand-foot drops, no more freaking fits for the second navigator, but the plane was shaking like an old man with the ague and it was impossible to do more than cling to the nearest stanchion and just watch as the pilot wrestled with the controls.

"Finally, and almost instantly, we just popped out into the eye, into the most beautiful sight. It reminded me of the Vatican, the semicircle of marble columns around the plaza before St. Peter's, but these columns extended all the way around, 360 degrees, and the convection currents had pushed them 50,000 feet into the sky. It was exquisite. The columns, the updrafts that came with the surrounding thunderstorm activity, formed a perfect circle. Winds were minimal in the eye, almost nonexistent. I remember sunlight coming in across the eye. Sometimes eyes are covered with layered cloud, but not with Dot. You could look straight down from the airplane to the ocean, 18,000 feet below. We could see the churning, the white caps on the sea."

Over the centuries, many have seen the eye of a hurricane and survived to tell of it, though generally from the bottom looking up. Sailors have reported stars at night, pristine blue skies by day. The air is reported to be luminous, with an unearthly gleam, with colors in a demented palette of lurid blues tinged with violet and somber greens. Winds can be utterly calm—unlike the water, which is churned by the surrounding and conflicting winds into raging and directionless mountains. Not all storms have eyes. Some that do are shaped like massive replicas of the old Roman Coliseum, sloping in from the top, perfectly round, as though they were forming seating for the gods, waiting for a gladiatorial combat of giants to begin below.

The sound in the eye is deep and ominous, like a freight train passing inches overhead, numbing the brain.

Max and his fellows kept their course, banked left. "We had to locate the center so we drew a line across, I watched the radar altimeter and marked the location of the lowest pressure, then cut back along a left turn. There was not much wind, only a slight descending motion. We measured our drift with a drift meter, our best way then of measuring wind speeds. As we made a ninety-degree turn, we released the dropsonde."

The adventure of Dot wasn't quite over. "We completed the mission and headed home. On the radio we heard that we couldn't land because there was a foot of water on the runway in Hawaii. The nearest alternate runway was Johnson Island, roughly 1,200 miles southwest from where we were. The navigator took a look at the distance, the course, and the fuel reserves, and said to me, 'We need some winds, to get us home.' I took the last maps I had, which were six hours old, and tried to put together a picture of what the wind pattern would be. I gave my winds to the navigator, and he sent them to the engineer. His job was to take my wind and convert it into 'fuel.' According to my calculations, we would be able to see Johnson Island just about the time the tanks emptied. On his own, the pilot decided to descend to see if he could find more favorable winds near the surface. We went down, to no more than two thousand or three thousand feet, and flew at this altitude for several hours.

"As I had predicted, just as we saw the lights on Johnson Island, the engineer announced that we were empty. We couldn't go around into the wind—no time. The pilot took a downwind leg and landed the plane, then reversed all four engines. We skidded the length of the runway."

After the plane stopped and the crew got out, Max took a look at the wings. Rivets had popped out all down their length.

Max penetrated six hurricanes in his time with the Air Force, between 1953 and 1958. Of all his reconnaissance flights, the most disturbing wind he experienced was not natural but manmade—he and his crew were assigned to fly into the mushroom cloud of a nuclear explosion. Why would they do this? "To see what it is like in there . . ." This was the 1950s. Radiation was known, but its larcenous carcinogenic qualities were not appreciated—scientists still regularly stood unprotected in the Nevada desert to watch the mushrooms ascending into the sky.

Max won't talk about the specifics of the mission—it was classified then, and still is. What he will say is this: The mission was to take readings inside the stem of the mushroom at 18,000 feet. Max was the radiological officer. "After detonation we went into the stem. It was the most beautiful thing I have ever seen, I saw shades of blacks, reds, pinks, and purples, colors that I never saw, ever again. The temperature was hot in the stem. We flew at 18,000 feet, took our readings, and came out on the other side looking like a flocked Christmas tree. We were 'hot' with radiation, but that wasn't the worst of it. The coral from the ocean floor, close to where the bomb had exploded, had been vaporized, and it crystallized on the plane as we flew through. It was a huge problem, bigger than the radiation. It made the plane too heavy. We were going down. I told the pilot to head for the rainstorm showing on the radar to attempt to wash off the weight of the coral. We did. And we landed safely, thankfully. But that plane was toast. It never flew again."[25]

One other recon flight flew into a hurricane in the 1950s that had picked up radiation from a nearby nuclear experiment. All that the crew were told afterward was to take a long shower.[26]

In fact, the danger posed to aircraft through flying into hurricanes is generally not substantial. This is because most of the wind in a hurricane is horizontal, and planes simply become part of the flow, just as they do when they hop a jet stream on their way across the Atlantic. It is the downdrafts that are the problem, and they are much more likely in thunderstorms than in hurricanes. Thunderstorms don't spin. They are an instability in the atmosphere that causes rapid overturning of air, and very rapid ascents and descents. But downdrafts are not unknown in hurricanes, which do spawn both tornadoes and thunderstorms—Max hit a downdraft on the periphery of Dot, and planes have, indeed, been known to drop a thousand feet in a few seconds inside hurricanes, just as Max described. But in thunderstorms such downdrafts are much more common, which is why pilots take a good deal of trouble to avoid such storms when they can. Most experienced travelers can report at least one such downdraft. My own score is two—one over the Congo, when my Zimbabwe-bound plane faced a wall of thunder cells it couldn't evade, and another on a flight to Tel Aviv, during which the coffee wagon and its flight attendant actually hit the ceiling, damaging both and causing panic among the passengers, most of whom (considering the destination) thought for a frightful second that a bomb had gone off.

That downdrafts don't happen very often in hurricanes doesn't detract from the bravery of the early pilots, who really didn't know what to expect, in terms of turbulence, or wind speed, or indeed rainfall.

Chris Fogarty, who has flown into several hurricanes in the Canadian Hurricane Centre's Convair 580 turboprops, says it is mostly experienced as light chop, "like driving a car along a pot-holed road." Hurricane Michael in 2000 was somewhat turbulent along its east side, and Juan, of 2002, pushed the aircraft around a bit. "There were times when my stomach felt it was still a couple of hundred feet below." His worst experience was trying to land in the aftermath of the flight into Tropical Storm Karen, when a low-level wind abruptly blew the plane several hundred feet sideways just above the runway; they had to abort and were diverted to Quebec City, more than six hundred miles away.[27]

When a U.S. Hurricane Hunter plane flew into Hurricane Hugo in 1989, it too encountered unexpected turbulence. Peter Black, of the National Hurricane Center's research division, re-counted that "when we got there the winds were over 200 knots and we just about lost the airplane. It was a really rough ride. An engine caught fire. I have a vivid recollection of seeing the flames shooting by my window. The turbulence was so severe that the fuel regulator had malfunctioned. They were able to douse the fire right away and feather the engine, but that meant we were in the worst part of the storm with only three engines."

Still, it was all for the best in the worst of all possible turbulences, as Black confesses: "In retrospect [we turned up a] really unique set of data. They showed for the first time what these smaller-scale meso-vortexes [little tornadoes drifting around the spinning eye-wall] look like that were contributing to the storm's deepening. We still don't know what role they play exactly. But we were able to identify that as a new entity, a new scale of motion in a hurricane vortex that we had never really documented before."[28]

The military, or at least the Air Force Reserve, still operates Re-con 53, the Hurricane Hunters, but most reconnaissance monitor-ing has been turned over to the Hurricane Research Program within the weather bureau. Most flights are now arranged through

NOAA's Aircraft Operations Center, based in MacDill Air Force Base in Florida, which tracks hurricanes, monitors air quality, surveys whale populations, measures snow cover, and performs many other tasks.[29]

In the 1950s, the same decade that his military and political masters blithely sent Max Kozak into the heart of a nuclear fireball, anything seemed possible to science. If they could tame the atom, why, hurricanes and tornadoes should be a snap. Weather control, the notion of bringing planetary weather under the benign direction of the disinterested, politically neutral weather bureau, was in the air, so to speak, a refugee from the hoariest tales of science fiction. In the old days, rainmakers did dances and made incantations and sacrificed fowl or lambs or, occasionally, their virgin daughters, to the weather gods to make things happen. Modern science was beyond that. You should only have to understand what makes the great storms tick to know how to make the ticking stop.

A sample of this thinking is from Irving Langmuir, a meteorologist who headed a cloud-seeding project known as Project Cirrus: "We need to know enormously more than we do at present about hurricanes . . . [but] I think that, with increased knowledge, we should be able to abolish the evil effects of these hurricanes."[30] Amateurs enthusiastically joined in the hunt for the magic technology that would do the trick, and more or less loony suggestions ranged from firing artillery through their tops to disrupt the rotation, to using giant fans to divert them, to flying hundreds of propeller aircraft against their rotation to unwind them, a device used to good effect by Superman in one of the Christopher Reeves movies to unwind time. Perhaps you could cool the ocean by towing into place massive Antarctic icebergs? Or by laying a cooling filament on the ocean surface? In Max Kozak's time, plans were mooted to see if a hurricane could be blown apart by a nuclear explosion; before it could be tested, wiser heads prevailed, suggesting that the added heat might actually intensify a hurricane, and in any case, what about the dispersal of the consequent radiation?

The notion of cloud-seeding, though, persisted well into the 1960s. It had been proposed as early as 1947, when scientists working on problems associated with aircraft de-icing found, for example, that moisture could be bled from clouds by a number of chemical reactions. Frozen CO_2, popularly known as dry ice, could turn water and ice into snow; silver iodide, for its part, could precipitate out supercooled droplets and cause rain. Both chemicals would have the effect of cooling down the furiously hot core of the hurricane, and perhaps calm it down. In the last few years of the 1940s, military planes ferried Project Cirrus scientists into the eyes of several storms, including one hurricane where they emptied canisters of silver iodide. To their delight, the hurricane began to disintegrate—but twelve hours later had reorganized and reenergized again. Worse, it had taken a sharp turn to the west, and ended up pounding Savannah, Georgia—the notion that they may have actually caused the turn made the Project Cirrus people blanch, even though they were assured it was a midlatitude ridge of high pressure that had really done the trick.

By the 1960s the National Hurricane Research Center understood more about hurricanes and how very powerful they really were. And how very big—masses of air and zones of high and low pressure reaching to 40,000 feet and stretching over thousands of miles. Still, Congress allocated $30 million to further the research, so further it they did, creating Project Stormfury to see if they could, once again, use technology to fatally disrupt a major storm. Two hurricanes were seeded with silver iodide, and indeed, maximum sustained winds dropped, if only for a brief period, by some 20 miles an hour, a not-insignificant result. It was tried again in 1969 with Hurricane Debbie. When the results were tabulated, seeding was seen to have some short-term, very short-lived, effects, but had exerted no real pressure and caused no real damage to the storm.

The extensive press coverage of the time persuaded both Cuba and Mexico that the United States was using weather-modification techniques as a form of ecological state terrorism; both countries demanded that the experiments cease forthwith. Soon afterward they did cease, though less because Mr. Castro demanded it than because the results didn't seem justified by the effort expended. But in the end the experiments were worthwhile. Much data was collected. Forecasters knew more. Hurricanes were better understood.[31]

Science still hasn't quite given up the idea of controlling the weather. An MIT scientist has bruited the equivalent of setting small controlled fires to stop major forest fires—in this case starting small tropical cyclones to head off larger ones by cooling the ocean and thus robbing hurricanes of their energy source. His unlikely scheme, presented to a weather-modification symposium early in 2005, involves a chain of offshore barges loaded with a series of upward facing jet engines. Each barge would create an updraft, causing water to evaporate from the ocean's surface and thus lowering its temperature. The resulting ministorms should dissipate harmlessly. The scientist Moshe Alamaro blithely put the cost at a billion or so a year. Also among the optimists is Ross Hoffman, a scientist at a company called Atmospheric and Environmental Research in Lexington, Massachusetts, who has been given a $575,000 grant from NASA's Institute for Advanced Concepts to look into the possibilities. Along with a number of other environmental scientists, Hoffman has been using Lorenz's discovery, but the other way around—not to see how far a storm would veer with minute changes in input data, but to see what input data would be needed to push a storm off course. Hoffman asserts that he has twice successfully "steered" storms, at least on his computer.[32]

Hurricanes are chaotic systems. But chaos doesn't necessarily mean what its commonsense definition implies—chaos has rules of its own. Chaos theory contains arcana such as strange attractors, which are only strange in that they cannot be analytically computed, though they are physically real, and can be readily observed. They are called attractors because "chaotic systems never wander aimlessly through the universe; they always hover around one of a finite number of dynamic forms."[33] If—a big *if*—we can track the loci of these strange attractors, we might be able to find the right sort of trigger for a hurricane, and if we then deploy it at the right moment, we might be able to turn storms away from where we don't want them to where they would do the least harm. Or, if we are not careful and the right moment turns out to be the wrong moment, to where they would do more harm than ever. Now there's an insurance company's nightmare.

Another idea bruited about was to coat the ocean beneath an embryo storm with biodegradable oil, to separate the disturbance from its fuel, the warm water of tropical oceans. The idea seems fine, but the problems are enormous—somewhat akin to the big bad wolf huffing and puffing at a fortress instead of a hut. Where to find enough oil, especially biodegradable oil, to spread over so vast an area? How to get enough vessels to the right place at the right time to do the spreading? How, indeed, to know which embryo disturbance was going to turn into a hurricane in the first place? And afterward, provided it worked, what to do about the wild creatures, the whales and the dolphins and the sea birds, affected by the oil?

On Hoffman's computer, a change in wind speeds of a mere 5 miles an hour was enough to steer a storm past an island instead of over it; and something as apparently small as a one-degree change in temperature at the initial stage killed a simulated storm completely.[34] But to change a thousand-square-mile piece of the atmosphere by one whole degree would take a massive amount of energy, almost as much as was latent in the storm itself. Where would this energy come from? Will satellites eventually beam down enough heat to change atmospheric temperatures? What unforeseen consequences will that have on weather and climate? Most ecologists believe the whole notion of control is misguided, and that man's natural penchant for meddling with the environment can only make things worse.

And there is one more difficulty with translating computer models into reality. Scientists can do hundreds of runs in the simulation models to find the one small tweaking that would cause just the right result. It can't be done by trial and error in the real world, because there is no way of telling ahead of time what consequence any single small change will have. The chances of making things worse are equally as good as making them better.

Despite all this, one fairly simple-to-execute method does exist to kill budding hurricanes while still tropical storms. If a massive explosion were to take place beneath the ocean, driving cool water from the deep to the surface, it would deprive the storm of its fuel and would stop it in its tracks. But an explosion that large would pretty much have to be a nuclear fusion device, a hydrogen bomb, and it's hard to see such a thing being tolerated by any population keen on self-preservation—any such explosion would cause extensive radiation contamination, massive fish kills, and tsunamis, which collectively would do more damage than the hurricane itself. Much better to follow the ecologists' dictum, and do nothing. Put the money to living more prudently, construct stronger and more hurricane-proof buildings farther inland, and leave the shore to the wild winds and the sea birds.

CHAPTER SEVEN

An Ill Wind

*I*van's story: By the morning of September 10,
Ivan had taken its expected turn from its westerly course to the northwest, and
most of the models now had it headed straight for Jamaica. After that, probably western Cuba and then to skim up the west coast of Florida. All day,
as Jamaica braced for the storm and battened down as well as it could, the
storm's intensity fluctuated. It started the day as a Category 5, but by mid-
morning the pressure had risen slightly to 929 millibars and the winds were
down to 125 knots or less, dropping the storm back to a 4. The deep pattern
on the north side was looking a little ragged, but no one thought the storm
was actually weakening, only changing. There was little wind shear to dis-
turb it, and the ocean water remained very warm. That life-saving ridge of
high pressure was showing some weakness over the Gulf of Mexico, and
Ivan would likely be steered around it to the west.

In the afternoon, as Ivan bore down on Jamaica, two things happened.
It reattained Category 5 status with the winds increasing to 165 miles an
hour, and the pressure dropped back to 910 millibars. This made it the
sixth most intense storm ever recorded in the Atlantic. But it also took
an unexpected jog back to the west from northwest—the high-pressure
ridge again—which meant that it struck only the western tip of Jamaica.
The eye and the strongest winds remained south and west of the island,
at sea.

Even so . . . The winds that did strike were the equivalent of Category

3 or slightly above, and the entire island lost its power and public water supply. Communities along the south coast were destroyed in the winds and the twenty-four hours of torrential rains that followed. The village of Portland Cottage and sections of Kingston were devastated. A dozen people were killed. Hospitals had to close, roads were torn up.

Millions of insects and thousands of birds, some of them endemic to Jamaica, were caught up in the violent spiraling winds and flung into the atmosphere. The animals were bruised and torn from their brutal evacuation, and remnants were found in later days across wide swathes of the Caribbean and as far away as Mexico, a graphic illustration of air in long-distance motion.

The storm passed thirty miles south of Grand Cayman. That was not nearly enough for safety. The smaller Cayman islands of Little Cayman and Cayman Brac were evacuated to the main island, but Jamaica had been reporting a storm surge of twenty feet or more, and the whole island of Grand Cayman was not much higher than that.

The scuba divers had all gone, flown out days before, sharing the evacuation planes with whatever holders of the Cayman's famous numbered accounts had been there on business, but the residents who remained reported the terrifying scenes—half the island under water, the airport vanished, one out of two of the island's 15,000 houses damaged (and there were no shantytowns on Grand Cayman, which has one of the strictest building codes in the Caribbean), roofs torn off, buildings collapsing . . . The Associated Press quoted banker Justin Uzzell, who was watching from his fifth-floor window before prudently taking refuge lower down, saying that "this is as bad as it can possibly get. It's a horizontal blizzard. The air is just foam." The winds tore the tops off the gigantic waves and drove them clear across the island, reducing visibility to near zero.

The storm was traveling west northwest at 10 miles an hour. The computer models' track estimates were slightly west of their previous day's forecast track, but they all still called for a curve toward the northwest and then north, through the weakness in the subtropical ridge, within forty-eight hours. This new track would take Ivan through west central Cuba, and then

*up the Gulf of Mexico parallel to the Florida coast, perhaps hitting the
Panhandle before exiting through Georgia.*

*There was no sign the storm was diminishing. On the contrary, the con-
ditions seemed conducive to a further strengthening, with warm surface temperatures and only light vertical shear. Ivan could easily reach Category 5
again before slamming into Cuba.*

*Its capriciousness began to seem intolerable to those people who were, or
could be, in its path.*

*Cuba had already evacuated more than a million people from the south
coast, and now authorities evacuated 300,000 more, moving them from the
western tip of the island farther inland.*

*The Cuban media had taken to calling the storm Ivan the Terrible. The
American media—or at least a large segment of the Florida media—insisted on
referring to Cuba in their weather news reports as "the communist-run island." Perhaps in retaliation, Cuba boasted of its people-friendly prepara-
tions. Indeed, they shut down the electricity grid some hours before the storm
arrived, thus preventing hundreds of transformer explosions and other electrical damage. Ivan, which seemed an equal-opportunity destroyer careless of human ideologies, just plowed onward.*

I

We have a narrow boardwalk down to our rocky beach—it is the
same one that Hurricane Juan pushed into the forest a few years
ago, now rebuilt—and at the beach end we built a small cedar
bench. One winter morning I spent an hour on the bench, watching the restless sea and a couple of harbor seals gliding through the
swells, their snouts and whiskers glistening in the pale sunlight. A
few eider ducks were splashing near the shore. The wind was somewhere, I guessed, between Beaufort 0 and Beaufort 1, really nothing
but a gentle onshore breeze coming out of the southeast. With the
glasses I could see a small boat out to sea. There, clearly, the winds
were more active, because the boat was rising and falling in the

swells and the horizon looked lumpy. Some kind of a front was coming in from the west, nothing very much, just enough to raise up a sea to our south, and this was the first sign of it. The breezes the front was pushing, which would bring us some rain, were part of a massive rolling wave of slowly tumbling air that stretched from Labrador well down into Pennsylvania, bringing damp but otherwise benign air all down the eastern seaboard. The jet stream was that week making a lazy loop almost down to the Gulf, and this loop was helping to steer the whole system. Behind the front—you could see it on the synoptic weather maps—was a massive plateau of still air all the way past the Great Lakes and almost to the prairies and south to the High Plains. West of that . . . It was snowing on the Sierras; another front was pushing the damp air out of the deep Pacific eastward, and it was dropping its load on Mammoth, where I have a good friend who even then would be digging out his SUV to get to the skiing trails—snow is a mixed blessing in Mammoth, but the good always outweighs the bad, except in a drought. He'd have another two feet of snow to deal with today, perhaps more. My front was connected to his, with only two lazy whorls of separation. Two whorls and more than three thousand miles. How many for that Pineapple Express that dumped snow on New England after starting somewhere down around Hawaii, and detouring past the Aleutians? Not more than four. Maybe only three.

Earlier, I had spoken to my mother, who was at a little beach community called Pringle Bay, an hour or so from Cape Town—she had gone to ground there with the rest of the family for Christmas, partly to escape the heat in a period of grim water shortages, and partly because, well, that's where the family went for Christmas, sons and daughters and cousins and a dozen children with assorted dogs, tumbled together in a kind of friendly chaos. The weather at Pringle Bay was hot and sunny—it was 36° Celsius, and they were having dinner outdoors. Their winds were southeast onshore breezes, not much stronger than ours, ten thousand miles away on the other side of the earth. I looked up the synoptic map for the southern hemisphere, courtesy of the Australian weather bureau's Web site, and saw a pattern not very different from the one we were experiencing. There were no typhoons in the Pacific, but there were several frowning waves of disturbed air, and the Furious Forties were as furious as ever, the winds scudding eastward in the midlatitudes between the Cape of Good Hope and Antarctica. Gales blew up whitewater just a few hundred miles south of Pringle, but the family had gone to the beach anyway, blissfully oblivious; the wind wasn't even strong enough for the beach sand to sting, as it often did.

How many degrees of separation between me and Cape Town? More than between me and California, perhaps, but surprisingly few, all in all. A picture came into my mind of the world as a big room, drafty, with maybe a window open (the poles) blowing cold air into the center, setting off corner-to-corner eddies and dips and whirls and swirls, with vortexes in the corners and in places where the cold and warm air collided. Were the air a thinly colored mist, you'd be able to see how it was moving about the whole room, and you'd be able to picture how completely the patterns would change if the window were closed and a door opened, say, or a fire was lit in the fireplace . . . But in the end the very homeliness of the metaphor didn't really work for me, because this was such a very big room with such very big drafts, and the equalizing dance of the winds was so intricate that it demanded a grander metaphor, a "Dance of the Seven Thousand Veils," if you will, the complex pavane of the global balancing system, and at that moment I felt that my small breeze, the one that was ruffling the feathers of the mergansers and eiders, was connected to the planetary whole in a way I hadn't really felt before.

This was not as comfortable a feeling as you might expect, because there is a practical downside as well as a philosophical upside to interconnectedness. We're all shut-ins in this great global ballroom.

We are locked in without a key, and there are more and more of us all the time, millions upon millions of us, and we are filling the air with our "smoaks" and our industrial defecations . . . Possibly this sour mood was brought on because I had been thinking about Ivan's recent and apparently malevolent and psychopathic presence in the southern seas, and at the same time had been contemplating the grim matter of air pollution, and had been trying with limited success to sort out fact from propaganda. The evening before I had been dipping into a score of books on the subject, the luminous Bill McKibben's *The End of Nature*, Donella Meadows' *Limits to Growth*, Jared Diamond's *Collapse*, Ronald Wright's *A Short History of Progress*, and a good deal of their gloominess had rubbed off—such a concatenation of woe! so many sins of commission and omission! such a certainty of calamities to come!—and for a moment our planet seemed less like a great global ballroom than it did an enormous ward for the criminally insane, and we the inmates, capering about and setting fires and wondering why the smoke was choking us. The mood didn't last long—my antidote was Richard Fortey's *The Earth*, a lovely look at the planet from the inside out, but still . . . As I had come to learn, Ivan was a part of the inherent balance of our planet, and in his violent way an inevitable and even positive force. We didn't create Ivan—he just *is*. But it is possible that through our "smoaks" we are creating the preconditions that make more and more awful Ivans probable. Pollution is not, after all, something we do in a vacuum, with no precedents and no consequences. What we do to the air can and does and will affect wind and weather and climate. This much seems obvious.

And no one seems to be in charge, much. Not in the world, and not in most of the major polluting countries.

The United States is an egregious case in point. Among the better guardians of American ecological purity is the Worldwatch Institute out of Washington, D.C., and among the rational voices that emerge there is that of Ed Ayres, the former editor of the *Worldwatch* journal. In a piece written in 2004, Ayres pointed out soberly that in the United States at least, no government agency exists to look after air as such. A multiplicity of agencies concern themselves with minute aspects of air—there are people looking at emissions, for example, and yet other people looking at emission controls, but no one looks after the whole. "The Environmental Protection Agency regulates some aspects of auto pollution, but the Department of Transportation regulates others, and the National Institutes of Health still others. You have to distinguish between people who regulate CO_2, and the people who regulate CO_2 emissions. Smog is a different department than global warming. Fuel efficiency is a different department than tailpipe emissions. Every component of the air had an agency responsible for it. But no one was responsible just for the air."[1]

Of course, Ayres is describing the classic reductionist thinking of the current state of Western science—the notion that by understanding something on the molecular level you can thereby understand its purpose. As he put it, "modern science's trend towards attempting to explain large phenomena as accumulations of tiny atomic or cellular ones misses the effects of the phenomena as whole systems." That is, we try to understand a tree by minutely examining its capillary and circulation systems and the molecular structure of its leaves, but we seem to have no appreciation for it just as a tree.

The people who study wind are doing better, I think, perhaps because wind is the most obvious part of air, and understanding a storm persuades hardly anyone anymore that we can control it. Atmospheric scientists, on the most theoretical level, have broken through some kind of limiting conceptual barrier: They are indeed delving deeper into the molecules, but have also regained a clear view of the whole global nature of wind systems. Perhaps this is because meteorologists, so constantly chastened by getting their forecasts wrong, have come to understand the virtues of humility.

Winds travel. We know this from the global models. Long-distance winds govern the health of our planet, but wind's long-distance travels aren't always benign, and some of the things we humans are doing to them makes their ill effects worse. It's not hard to collect examples. It is harder, in fact, to ignore examples, since so many of them turn up in news reports, most of them in some way "true," though often reported out of context.

In the late summer of 2002 I had been filming a documentary about water in north China, on the fringes of the Gobi, and had noticed that the air overhead seemed curiously opaque, even milky; there was no blue, even on clear days. This was more evidence of China's inexorable desertification and consequent dust storms. Alas, Chinese efforts to correct the problem may be making it worse—in mandating that marginal land must be used for farmland, the government simply encouraged practices that caused the newly plowed soil to blow away. In 2001 NASA had tracked a massive dust storm originating in north China big enough to briefly darken skies and cause hazy sunsets over North America as little as five days later; a month after I left to return home, NASA's satellites once again picked up an immense dust cloud, more than a mile thick, moving eastward over Korea and into the Pacific. Clouds like it seemed to be becoming part of the Chinese calendar. Similarly massive dust clouds had occurred in 1997, 1998, and again in 2000; in fact, the Chinese Meteorological Agency counted twenty-three major dust storms in the 1990s, a substantial increase over previous decades. Chinese dust—heavily tainted by pollutants such as coal-combustion aerosols, ozone, persistent organic pollutants (POPs), and heavy metals such as mercury—has inflicted itself on Korea and Japan for decades; in Korea it is sometimes called spring's gatecrasher. Chinese dust may have been the origin of an outbreak of foot-and-mouth disease on Korea's west coast.

The same summer, 2002, a United Nations Environmental Program (UNEP) study confirmed the existence of another pollution cloud, two miles thick, over much of southern Asia. Klaus Töpfer of UNEP said in Vijay Vaitheeswaran's *Power to the People*: "The haze is the result of forest fires, the burning of agricultural waste, dramatic increases in the burning of fossil fuels in vehicles, industries and power stations and emissions from millions of inefficient cookers burning wood, cow dung and other biofuels . . . There are also global implications—not least because a parcel like this can travel halfway around the globe in a week." Every year in developing countries, at least a million people die from outdoor air pollution. "Disaster is not something for which the poorest have to wait; it is a frequent occurrence," Cambridge University professor Partha Dasgupta is quoted as saying in *Power to the People*.[2]

Indian researchers have studied the high concentrations of black carbon (a.k.a. soot) over the Indian Ocean, and traced it back to biofuels, mostly cattle dung, used for cooking fires by millions of people in the subcontinent. Only by changing the way India cooks, the study suggested, could the country help mitigate climate change. They acknowledged that change was less than likely.[3]

Even such mundane human byproducts as dandruff have been found in the pollution clouds. A study in 2005 found that "particles injected directly from the biosphere" are a major component of atmospheric aerosols. Examples given were fur fibers, dandruff, skin fragments, plant fragments, pollen, spores, bacteria, algae, viruses, and protein crystals. All these, the study suggested, have a substantial impact on climate through cloud formation—they attract water and make excellent ice nuclei, which triggers rainfall and removes water from the atmosphere.[4]

A 2003 U.N. study on the North American environment, which tiresomely predicted, as these reports usually do, ever more droughts, floods, and severe storms (predictions based on very thin evidence), also pointed out correctly that total energy use on the continent grew 31 percent from 1971 to 1997; that 5.5 million people have developed asthma and other severe respiratory diseases as a result of air pollution; that the Ontario Medical Association's figures put the air pollution fatalities every year in Ontario alone at 1,900 people, costing the system well over a billion dollars.[5]

At Tagish in the Canadian Rockies a monitoring station has found elevated levels of pesticides in the winter and spring, attributed to pollution from continental Asia. Similarly deleterious effects have been noticed in focused studies of snow cover in Alaska and British Columbia and on the fecundity of Pacific eagles. A study on the Fraser River watershed in British Columbia concluded that toxic airborne pollutants from Asia have been contaminating lake fish and sediments; high POP concentrations have been found in the snowpack. Farther south, increased nitrates and sulfates have been detected in pristine streams in the Olympic National Forest on the coast of Washington State. Other studies document POPs and mercury in wildlife and human populations in the Arctic, pesticides in bald eagles of the Aleutian Archipelago, and very high polychlorinated biphenyl (PCB) concentrations in some Pacific Northwest orca populations.[6]

Asia is hardly the only villain—villains are to be found wherever winds blow. In late 2004, for example, Arkansas soybean farmers were lamenting the ravages of soybean rust, a fungus that had landed in the United States, blown from South America and carried ashore over the Gulf of Mexico by one or another of the season's hurricanes. The fungus was also attacking another alien import, kudzu, which state governments had been vainly fighting for decades, so many people who were not growing soy could see the bright side.[7]

And I had seen for myself how strong winds picked up massive clouds of African dust, and carried them out over the Atlantic. This, too, was hardly new: For decades it has been known that pre-Columbian pottery in the Bahamas was made from wind-borne deposits of African clay; orchids and other epiphytes growing in the rain forest canopy of the Amazon depend on African dust for a large share of their nutrients. Charles Darwin's *Beagle* journals contain an observation he made as he was crossing the Atlantic about the falling of "impalpably fine dust" on the ship at sea.

In the last year of the millennium a reddish brown river of dust, picked up from the deserts and the eroding grazing lands of the Sahel, a plume hundreds of miles wide and thousands long, was whipped across the Atlantic by the trade winds. The amount of transported dust has been going up steadily over the past twenty-five years, and at the same time, the mortality rate of creatures like Caribbean coral has risen sharply. Eugene Shinn, a researcher with the U.S. Geological Survey in St. Petersburg, Florida, has tracked the coral's declining health to fungal spores and bacterial cysts hitching a ride on African sand; in 1998, scientists identified an African soil fungus as the cause of the decimation of sea fans across the entire Caribbean, an object lesson in the interconnectedness of life. The red sunrises in Miami are Saharan-caused; half the particulates landing on Florida are from the Sahara. Dust clouds increase in Caracas when drought in the Sahel occurs—another example of the intimate links that winds make.

The same thing has been happening in the U.S. Virgin Islands, where the coral reefs have been dying for years. Most of the blame had been attributed to overfishing and to direct damage by boats and divers; but in 2000 several studies found that hurricane-carried pathogens from Africa had severely degraded vegetation and had critically damaged once-dominant corals like staghorn and elkhorn, long-spine sea urchins, and sea fans. Carpets of algae now dominate many reefs. The Virgin Islands National Park even paid for a marine ecologist to visit Bamako and Timbuktu on the Niger River in Mali in an effort to understand the organisms that were making their way across the sea. Similarly, another U.S. Geological Survey report in 2001 said that what the researchers called opportunistic pathogens were hitching rides from Africa on the wind—the sand is heavy enough that the dust clouds block the solar radiation that would otherwise damage the bacteria on their journey to the New World. Large dust arrivals from Africa have now been found over 30 percent of the continental United States; although no one has yet estimated its mass, it would be a small fraction of the amount that leaves the Sahara. About half the volume that reaches the United States settles on Florida. On any given day, a third to a half of the dust drifting through Miami comes not from local beaches but from Africa. "It may," the study suggested, "pose a significant public health threat."

This might seem something of an exaggeration, but in the summer of 2001 a NASA-funded study tracked a cloud of Saharan dust to the Gulf of Mexico, where it settled, with unnerving consequences— causing a huge bloom of toxic red tide. The Saharan dust reached the West Florida shelf around July 1, increasing iron concentrations in the surface waters by 300 percent. Through a complex process involving enzymes and plantlike bacteria called *Trichodesmium*, the iron enriched the nitrogen content of the ocean, and in October an 8,100-square-mile bloom of red algae had formed between Tampa Bay and Fort Myers, Florida. Red tides give off toxins that can cause respiratory disorders in humans, and also poison local shellfish. Anyone eating the contaminated shellfish would suffer paralysis and severe memory problems. This particular red tide also killed millions of fish and hundreds of manatees.

A study at Harvard in 2004 found an increase in dangerous epidemic diseases created by the decrease in global forests and by increasing numbers of devastating storms like hurricanes.[8] And it is known that these dust events bring chlordane and DDT traces back across the Atlantic, chemicals invented in, but now banned in, North America.[9] There's a nice irony: America is being bombarded with DDT from Africa and chlordane and lindane from Asia, very toxic chickens flying home to roost.

Pollution goes where the winds take it. Researchers from the University of California at Davis were monitoring the air quality on Mauna Loa, a 13,680-foot mountain in Hawaii, and to their dismay found clear traces of industrial pollution from China, including arsenic, copper, and zinc, kicked into the atmosphere five days earlier. "It seems that Hawaii is like a suburb of Beijing," one of them said. On the other hand, for Europeans, it is the United States who are the aggressors, and dirty air from the United States regularly fouls northern European forests. Beltway commuters in Washington, intent only on the politics of the moment, are actually damaging the lungs of hikers in Britain's Lake District. Aerosols including toxic metals, nutrients, viruses, and fungi have been tracked from the Gobi Desert to Beijing, from West Africa to the Caribbean, from Ontario to New England, and from Germany to Sweden.[10]

Science magazine in April 2000 published a study that attributed a sudden increase of carbon dioxide concentrations in the American Southwest to Canadian forest fires. Episodic spikes in emissions on the central Atlantic Coast of America were caused by forest fires burning thousands of miles away, in the sub-Arctic Northwest Territories. A single fire in this remote region emitted two and a half tons of CO_2 per hectare (about two acres) of forest burned, and dumped it on the United States—and total emissions from forest fires exceeded all other sources by a factor of two. This is unfair—Kentucky makes enough of its own pollutants without inadvertently importing others from Canada, but turnabout is fair play, I guess, for it is not for nothing that Nova Scotia is sometimes referred to as the tailpipe of North America, since sulfur dioxide and carbon dioxide from Illinois and other places in the Midwest tend to drift over—and sometimes onto—our heads.[11]

In the summer of 2004 the first really massive study of transoceanic pollution was undertaken, fronted by NASA and NOAA, with assists from Environment Canada, the British environment ministry, and scientists from France, Germany, and Portugal. This was the International Consortium for Atmospheric Research on Transport and Transformation, more usefully known as ICARTT.

There were some security hiccups related to sensitivities about high-flying foreign aircraft penetrating national airspaces, particularly on the part of the ever-prickly French, but to no one's surprise, the best scientific guesses were amply and miserably confirmed; high-altitude solar radiation was turning pollutants into lung-irritating ozone. Yes, Asian pollution was darkening the prospects for Californians and Washingtonians; but yes, also, U.S. pollution was being sucked into the westerlies and the jet stream, to be deposited in Europe three to five days later—those Beltway commuters were indeed affecting Britain. One of the British scientists involved, Alastair Lewis of the University of York, said gloomily that "we used to think air pollution was a local problem. Now we realize some pollutants, particularly ozone, are global. It is literally arriving here on the wind." Environment Canada's Richard Leaitch, for his part, was concentrating on how clouds process trace gases and particulate matter, but did confirm that pollutants from the United States were tracking northwest to the Maritime Provinces—the tailpipe didn't end in Maritime Canada, but it did leak substantially there.

ICARTT was the largest, but not the only, study of wind-borne pollution being carried out by atmospheric scientists in the early years of the millennium. Alarming data from all around the globe on aerosols' damaging effects on regional and global climate kicked atmospheric scientists into high gear, and by 2004 a bewildering variety of studies with impenetrable acronyms were being conducted. They variously measured ozone concentrations; sodium dioxide, carbon dioxide, and formaldehyde emissions; the worldwide spread of carbon monoxide pollution; and the global distribution of man-made and natural aerosols.

Main conclusions from all this frenetic activity? The good news was that stringent air quality controls in Europe had decreased sulfur pollution substantially. But that was pretty much it for good news, at least as reported. Even sulfur dioxide hadn't decreased globally, because a wash of pollution from Asia was more than compensating for the decrease in Europe. For the rest . . . global winds carrying ozone and carbon monoxide (CO) were jeopardizing agricultural and natural ecosystems worldwide, and having a strong impact on climate. All the studies recognized what was already obvious, that Asian pollutants were beginning to surpass those from North America, a trend that would only continue and accelerate.[12] The increasing concentration of carbon monoxide in the atmosphere worldwide was particularly worrying. The importance of megacities, defined as cities with more than ten million inhabitants, was recognized as a new and critical source of pollutants, especially from burning fuels—by 2001 there were seventeen megacities worldwide. And a final point: Pollution from elsewhere was making local conditions worse, pretty well everywhere.[13]

Well, they "knew" all this before. But now they know it for sure.

II

In no aspect of atmospheric sciences, of the study of winds and the air, is the discussion more heated than on the topic of the greenhouse effect and the growing presence in the air of carbon dioxide. On no other topic are the doomsayers more strident, and the doom they foresee more self-evident; and on no other topic is the cheery optimism of the naysayers more chilling. Here are two utterly typical quotes from my notebooks: "If humanity indeed adds another 200 to 600 parts per million to atmospheric carbon [which it is in train to do], all kinds of terrible things could happen, and the universe of terrible things is so large that some of them probably will." This quote is by Steve Pacala, a Princeton University ecologist. The other followed a discussion about the immense reservoirs of organic carbon stored away in the permafrost in Arctic regions—perhaps two hundred billion tons of the stuff, "safely" stored for thousands of years, because frozen. "Now," says Terry Chapin of the University of

Alaska, "it's potentially a very large time bomb." If the Arctic thaws, rising sea levels will be the least of our worries, in this view. Instead, we'll be smothered in CO_2 and in short order the planet will become as steamy as Venus, ending life as we know it, and most important, our own.[14]

This is bad enough, but perhaps the most depressing thought of all is to contemplate not so much what is happening but how rapidly it might happen. The potential time line is unnervingly short. The global climate, like its component parts, is a chaotic system with its own strange attractors. Theory suggests that there might be at least three such attractors. One is the current climate model, another is a White Earth model (the deep freeze of an ice age), and a third is the Venus model (with dense clouds and surface temperatures high enough to evaporate the oceans). If climate is indeed chaotic, it would tend to hover around one of these three, and have the ability to bounce unpredictably from one to another. If we are close to such a bounce, a very small effort on our part, such as an increase of a few parts per million of CO_2, might be a sufficient trigger. Unlike conventional global warming theories, these bounces wouldn't happen in a reassuring one hundred or two hundred years. They could be complete within a decade. We would have no time to prepare. Flip! The Venus effect . . . Or flip! The deep freeze . . . Fry or freeze. An evil choice.

The greenhouse effect is simple enough to understand, though not quite as simple as in the popular imagination. It is really the tipping point at which the air will become dangerously oversaturated with moisture vapor and carbon dioxide that is causing all the fuss. What concentration of CO_2 is too much? Where's the true danger point?

In the present mix of atmospheric gases, a little more than half the sun's energy reaching the outer atmosphere strikes the surface directly. The other half is distributed through scattering, "bouncing" off other molecules in the same way odor molecules disperse themselves in the air. Much of the energy that does reach the surface directly is absorbed, but it is then reradiated upward again. The reradiated energy has much longer wavelengths than solar energy, somewhere between 1 and 30 micrometers. Between them, carbon dioxide and water vapor absorb radiation at these wavelengths efficiently, except for a small window that is transparent to radiation, which lies between 8 and 11 micrometers. It is through this window that some of the reradiated heat is able to escape back into space.

The half of the sun's energy that doesn't reach the surface is absorbed by the same two substances, water vapor and carbon dioxide. The CO_2 molecules become agitated and therefore warmer by the process of absorption, and they then reradiate the energy they took in, some of it down to the surface, some out to space. That energy reaching the surface is treated the same way as the other half—it is absorbed and then reradiated spaceward. Some of it is trapped yet again and sent back down . . . and so on and so on, setting up an oscillating feedback effect not unlike a game of Ping-Pong.

The net effect of all this is that the atmosphere has been gradually heated to a relatively constant temperature with a relatively constant variation by altitude, somewhere around 6.5° Celsius for every 3,300 feet of altitude. It is this constant feedback effect that makes the proportion of carbon dioxide in the air so important—the more there is, the more absorption there is, and the greater the heat gain in the atmosphere.[15] This is pretty straightforward, complicated only by the fact that other greenhouse gases, like methane produced by cattle, rice fields, and landfills, and chlorofluorocarbons emitted from refrigerators and air conditioners, do the same thing carbon dioxide does.

No one, not even the most vociferous climate-change skeptic, quarrels with this analysis. Nor does anyone disagree that CO_2 levels are elevated over historic norms. CO_2 levels held pretty steady, around 280 parts per million, for the thousand years before 1800.

Since then, as industrialization really got under way, atmospheric concentrations of CO_2 began to rise. Today they are around 370 parts per million.[16] Some people think this isn't very much—after all, CO_2 is only a very small component of air, just a trace gas really, and an increase of 90 parts per million, if visualized as distance, would be less than a third of an inch in a hundred yards. On the other hand, it does represent a 30 percent increase, and most earth scientists believe that with the still-expanding fossil fuel infrastructure, it may be impractical to avoid 440 parts per million, a significant increase from present levels. It also seems to be agreed that each resident of the developed world adds around five tons of carbon to the air every year, most of it coming from exhaust pipes and smokestacks. There is less agreement about whether this increasing CO_2 has already led to global warming, and the most commonly cited "hockey stick" shaped chart, which shows a sharp upturn in global temperatures at about the time of the industrial revolution, is still widely disputed as being the result of statistical errors; there is in fact considerable evidence that parts of the world, including Europe, have gotten cooler instead of warmer over the same period. On the other hand—and there always seems to be another hand in this debate—a careful study by James Hansen of NASA's Institute for Space Studies and other scientists has concluded that the earth's "energy imbalance," the net heat gain over heat loss, is almost one watt per square meter of earth surface (enough, the authors say, that if it were maintained for ten thousand years, it would be enough to boil the oceans). Gloomily, they go on to suggest that rapid climate change might take less than a century, while it would take at least a century to change our ways even if we started now, "implying the possibility of a system [already] out of our control."[17]

Clouding the debate and fuzzying up its conclusions is the role of perfectly natural short-term phenomena such as, well, clouds.

In May 2004 NASA published a study of the earth's albedo, how much light and heat the earth radiates back into space, instead of absorbing, based on differences in observed earthshine on the moon. The data showed a steady decline, or dimming, of the earth's albedo from 1984 to 2000, with a particularly sharp decrease in 1995 and 1996. From 1997 to 2000, the earth continued to dim, albeit more slowly. Less heat going out to space meant more heat on the surface, and this correlated with the observed data—there was an increase in mean global temperatures in the same period. But in the past three years, the trend has apparently reversed, and the earth now appears to be both getting more sunshine and radiating more light back into space. "Though not fully understood," the study said with some understatement, "the shift may indicate [nothing more than] a natural variability of clouds, which can reflect the sun's heat and light away from earth. The apparent change in the amount of sunlight reaching earth in the 1980s and 1990s is comparable to taking the effects of greenhouse gas warming since 1850 and doubling them. Increased reflectance since 2001 suggests change of a similar magnitude in the other direction."[18]

Which means that natural cycles in cloud cover can account for changing surface temperatures; changing temperatures affect winds and wind patterns; winds create weather; weather in turn affects cloud cover . . . If taken literally and narrowly, the study's results could also mean that the relatively trivial amount of global warming recorded so far is caused by nothing more than changing cloud cover, and is climatically meaningless. The U.N.'s Intergovernmental Panel on Climate Change (IPCC) has basically admitted this might be true: "Clouds represent a significant source of potential error in climate simulations," its 2001 report said. Cloud formation might even have a negative feedback effect on atmospheric warming—that is, it might dampen warming in a way analogous to how the human iris shrinks when the light gets too bright. If true, the net effect would be that the predictions of global warming are highly exaggerated. (The conclusions of this study were disputed in early 2005 by another study, also led by a NASA scientist.)[19]

The most ardent disagreements among scientists, though, are still about what concentrations of CO_2 represent a real danger to life.

One of the main areas of debate is the role of the oceans. Oceanic CO_2 is rising as quickly as atmospheric CO_2. An NOAA study estimated that the oceans have absorbed 120 billion tons of carbon in the last century, most of it generated by the burning of coal, oil, and gas. The current rate of absorption is 20 to 25 million tons of CO_2 a day—a rate not seen on this planet for twenty million years. The accumulation is one hundred times faster than between the last two ice ages.[20] The oceans, then, are acting as carbon sinks—each year humanity pours somewhere around 8 billion tons of CO_2 into the atmosphere, but less than half of it stays there. The rest goes into the oceans.

One of the many things complicating the inquiry is the relationship between carbon dioxide, the principal greenhouse gas, and sulfur dioxide, a common pollutant. If—a very big *if*—we act to reduce CO_2, this will in theory slow the global warming trend. But because we are concurrently reducing SO_2, which will itself have a slight warming effect, the results could be masked. "There are so many similarly fuzzy factors—ranging from aerosol particles to clouds of cosmic radiation—that many parts of the world could endure unfamiliar weather patterns and maybe even freakish storms for years without knowing how it is happening or what to do about it," says Vijay Vaitheeswaran.[21]

Some of the CO_2 in the air and the sea is absorbed naturally. Mollusks, for example, take it from the oceans to make their shells, so farming mussels must be a good thing. On land, forests take CO_2 from the air to make wood; measurements taken in the air over large forests show that the CO_2 concentrations are ten parts per million lower than elsewhere. However, the optimistic theory that increased carbon levels will make forests grow faster seems to be wrong, because trees more quickly run out of other essential nutrients. Ironically, cutting down old-growth forests, a major concern of the

Green movement, might actually help—young growing trees need more carbon than older ones.[22]

In February 2005 a study found that humans are indeed warming the oceans, down to thousands of meters, almost certainly due to increased carbon dioxide caused by the burning of fossil fuels and the consequent greenhouse effect. The study's lead author was Tim Barnett of Scripps Institution of Oceanography in La Jolla, California. "This should wipe out most of the uncertainty about the reality of global warming," he asserted.[23]

All this connects, of course, directly to winds and to weather, albeit in ways difficult to untangle. If global warming does indeed cause the ocean temperatures to rise, and hurricanes need warm water . . . does it follow that there will be more, and more severe, hurricanes? It would seem so if you believe the news. In fact, if you follow the news even cursorily, you'll see that it's been a given for some years that global warming will produce more severe weather more often. It's an assumption that seems to have been generally accepted, even by many experts. It may even be true. But it isn't necessarily true.

The IPCC's 2001 report stated that there was no evidence tropical cyclones had increased in intensity or number, although there was evidence that the top one thousand feet of ocean had warmed half a degree, and said that the panel had no way of judging future trends—there simply wasn't enough evidence either way, and certainly not enough to support the popular contention that extreme weather was happening more often. It's fair to say, though, that this conclusion was not at all unanimous. One of the leading authors of the IPCC's next report (scheduled for 2007) was quoted in 2005 as suggesting that warming sea temperatures and rising sea levels caused by global warming were changing conditions for hurricanes, and that the extrabusy 2004 and 2005 hurricane seasons may well be a harbinger for the future. He was promptly accused of politicizing the science, and a flurry of resignations from the IPCC followed. The author retorted that he hadn't meant there'd be *more* hurricanes, only *bigger* ones, but this gloss mollified no one.

Obviously, temperature changes in the oceans have some effect, but what? Some hints can be found, but they are tantalizingly vague. For example, the southern Saharan fringe was unusually damp in the 1950s, and suffered unusually devastating droughts in the 1970s and 1980s. In 2004, as acknowledged in the first chapter, it rained again along the southern Sahara, which includes Darfur. These rainfall data correlate neatly with temperature shifts in the oceans, whether higher than normal temperatures in the southern Atlantic or Indian oceans, or lower than normal temperatures in the North Atlantic, which in turn neatly correlate with the waxing and waning of hurricane cycles. A one-on-one causality is far from proven, but the coincidences are startling.

In 2005, near the end of a uniquely busy hurricane season—the season in which Katrina, Rita, and Wilma all battered the U.S. mainland—*Science* published a new statistical study that actually showed the number of storm days and the raw number of tropical cyclones decreasing over the past decade, in all ocean basins except the Atlantic. The proportion of severe storms, Category 4 and Category 5, however, increased sharply.[24]

This still doesn't tell us, though, whether severe weather will increase. Or decrease. El Niños, as we have seen, can sometimes have a mitigating effect. Similarly, hurricanes affect climate, just as climate affects hurricanes. But the global General Circulation Models (GCMs) aren't consistent. Just as some say the American Great Lakes will dry up as the world warms, and others speculate that they might actually increase in volume, so storm predictions are all over the map, quite literally. Climate change could raise upper-level atmospheric temperatures, or it could lower them, which would increase or decrease the differentials between surface and high-level temperatures, and so change the threshold point for hurricanes. Also, a warmer world might have stronger upper-level winds, which would kill hurricanes as strong upper winds do now. Or it might increase El Niños, which would increase Pacific typhoons but decease Atlantic hurricanes. Or none of the above. Severe weather might even become less frequent in a warmer world, not more so. If you see a confident prediction about the bad stuff coming because of global warming, treat it with the utmost skepticism.[25]

III

Because ours is a technological age, dozens of high-tech solutions have been proposed for solving the atmospheric carbon problem, to the evident disdain of ecologists, whose solution seems mostly to be to leave the carbon where it is, in coal mines and oil fields, and cut down consumption instead. Some of these solutions might work; others seems reminiscent of the scheme to control hurricanes by flying propeller-driven aircraft into them to unwind their rotation.

An example is the grove of rotating three-hundred-foot synthetic trees proposed by scientists at the Los Alamos National Laboratory in New Mexico. The notion was to use giant plastic blades to direct wind onto a filter dusted with sodium hydroxide. The resulting by-product, sodium carbonate, would be scraped out and heated to free the CO_2, which would then be compressed for storage. Jennifer Kahn, who reported on this and other outré storage schemes in a piece for *Harper's* magazine in May 2004, suggested sardonically that the trees would function like a kind of atmospheric kitty litter. The Los Alamos press release that announced this project claimed that a mere twenty thousand of these ghastly trees would be enough to absorb all the CO_2 from all the cars in the United States. Why stop there? "Cover the entire state of Arizona," Kahn suggested, "and there would be theoretically enough for all the cars in the world." Another lunatic scheme she uncovered was proposed by Craig Venter, the man who helped sequence the human genome. His group wants to create a synthetic microbe that will eat CO_2 and excrete it as fuel. By 2004 they had already found several natural bacteria that consume CO_2 and convert it into methane and hydrogen, and were looking to increase their efficiency. As Kahn said, laying on the irony pretty thick, "a policy that promotes the burning of irreplaceable resources at the expense of the climate makes perfect sense [in] a world not of humans but of posthumans, who eat carbon dioxide and shit coal."[26]

Another scheme would be to boost the oceans' appetite for carbon by sprinkling the sea's surface with billions of iron filings. Millions of acres of marine algae are, in effect, anemic. As *National Geographic*, which reported on this scheme, put it, what stops them from absorbing much more carbon is lack of iron, hence the "Geritol solution."[27] Or perhaps we could just take the carbon out of coal. It could be done too: Coal could be induced to react with oxygen and water vapor to make pure hydrogen, plus waste gases including carbon dioxide, which could then be buried underground.

Burial, now given the grander word *sequestration*, was by 2005 the most popular of all the suggested technological remedies. It's a perverse twist on the ecologists' notion of leaving the carbon where it is—take it out, use it, then put it back.

In September 2004 the Toronto *Globe and Mail* reported on what it called a major four-year study that suggested something miraculous: The oil industry could squeeze more oil from almost depleted fields and, at the same time, take care of at least some of the carbon problem. That this seemed suspiciously neat, and the fact that a major Canadian oil producer, EnCana, was involved in the study, apparently gave the business writer no pause for thought. In any case the study was presented to an international conference of Greenhouse Gas Control Technologies by Malcolm Wilson, director of energy and environment at the University of Regina. The *Globe and Mail* quoted him as saying: "This wasn't a small pilot test, or simulated results. We're doing big tests in a real world environment."[28]

What happens is this: Carbon dioxide gas is injected into an oil well to mix with the remaining oil, making the oil less viscous, enabling it to be drawn more easily to the surface. The carbon dioxide, for its part, stays where it is put. The researchers used the opportunity to test the long-term storage of carbon dioxide. They concluded the gas can safely be stored in old oil reservoirs, "although further work must be done before greater certainty can be attained on a longer-term scale, say over hundreds of years." For its part, EnCana said the study "is an example of how oil production can be increased while helping the environment." Ironically, EnCana didn't have enough CO_2 of its own for the test, and had to import it from North Dakota, although Canada, a signatory to the Kyoto treaty, had seen its carbon dioxide emissions rise 13 percent between 2004 and the time the treaty was mooted, whereas the United States, which refused to sign, increased emissions by only 7 percent in the same period. (U.S. emissions in 2003 were actually below 2000 levels.)[29]

The United States has allocated some $110 million for sequestration research. But even if it shows good results, the old wells and mines will eventually fill up, so researchers are already contemplating using the ocean to sequester the gas. Jennifer Kahn again: "Some scientists envision running pipes from the flues of seaside refineries pouring CO_2 towards the ocean floor like bubbles through colossal straws. Others imagine an even more ambitious scenario, in which CO_2 could be pumped so far down it would emerge as a hydrate, an ice-like solid."[30]

The British government has already stored millions of tons of CO_2 in depleted oil wells under the North Sea. They have been there for several years. So far, no problems have been reported. But several years, as environmentalists will insist on pointing out, are not the thousands that would be needed.

So the air is getting fouler every year, and the ticking time bomb of carbon dioxide and greenhouse gases is set to go off any second? Well, yes, but that's not the only news. Though you would be hard put to hear it over the din of the breast-beating and lamentation, there is also some good news, rather more good news than the plethora of international studies cited earlier has acknowledged. Europe, as we have seen, has so reduced its emissions of sulfur dioxide, one of the worst aspects of coal burning, that even the most rigorous scrutineers among the Greens have declared themselves impressed. In 2004 the European Parliament voted for tougher standards on pollution caused by heavy metals such as arsenic, cadmium, nickel, and mercury, and by polycyclic aromatic hydrocarbons, or PAHs, a group of over one hundred chemicals that are formed during the incomplete burning of coal, oil, gas, garbage, or even charbroiled meat or tobacco. In the United States, a national snapshot of air quality shows improvement for almost every kind of pollution—"with particularly dramatic declines in carbon monoxide, sulfur and lead. Between 1976 and 1997 levels of all six major pollutants decreased significantly, sulfur dioxide levels by 58 percent, nitrogen dioxides decreased 27 percent, ozone decreased 30 percent, carbon monoxide decreased 61 percent, and lead by 97 percent."[31] But a new report published in 2004 by the National Research Council, while agreeing that significant progress has been made, particularly in the emissions that led to acid rain (mostly various sulfates), and while declaring that the Clean Air Act has actually achieved its purpose, nevertheless warned that "many areas are [still] not in compliance for ozone and particulate matter" and called on the national government to do more to cut emissions from older power plants, diesel trucks, and nonroad diesel engines.[32] And the study did warn that more attention should be paid to the long-distance carriage of pollutants on the global winds.

Another interesting indicator of progress was a piece in *Science* questioning the notion of attempting to move to hydrogen-based fuels for cars, not because they are intrinsically a bad idea, but because "regulation-driven technological innovation has reduced emissions from gasoline-powered cars to the point where they have very low emissions per-unit-energy compared with other sectors and other transportation modes. This trend will continue, reducing the benefit of zero-emission hydrogen vehicles, particularly because many technologies (e.g., electric drive) can be used on both platforms."[33] Despite these reservations, Iceland opened the world's first hydrogen filling station for cars in April 2003, and has announced plans to become the first completely hydrogen-based economy, fossil-fuel-free, by midcentury. And even the born-again former Hummer driver, Arnold Schwarzenegger, signed an executive order mandating hydrogen filling stations for every twenty miles along California's network of state-owned freeways.

Big Oil, too, has lumbered cautiously onto the bandwagon. Shell's former boss thinks the Kyoto Protocol is crucial because it forces businesses to put their best and sharpest minds on the task of reducing carbon emissions. Exxon, a notorious scoffer about the "ludicrous junk science" of global warming, is actually investing huge sums in energy efficiency, geological sequestration, and other low-carbon technologies as a way to hedge its bets.[34]

Then, in Davos in February 2004, eleven very large companies, major polluters all, made a commitment that their activities would be laid open for all to see—they promised to disclose and detail all the greenhouse gases they produce on a new open Web site called the Global Greenhouse Gas Register. The register was launched to considerable media fanfare by the World Economic Forum; the pious declaration that accompanied it hoped that other major companies would follow their worthy lead.[35] Together these companies account for some eight hundred million tons of CO_2 a year, fully 5 percent of the total emitted by the thirty-seven industrialized nations governed by the Kyoto Protocol. The companies also promised to prepare corporate-wide inventories of their other major greenhouse gas emissions—methane (CH_4), nitrous dioxide (N_2O), hydrofluorocarbons (HFCs), perfluorocarbons (PFCs), and sulphur hexafluoride (SF_6)—and to have had, or be prepared to have, that information independently verified. By year's end the site was still largely empty—only two companies had reported in, and the data available to the public were, to put it mildly, sketchy. Still, the idea was a good start.

The following year, in April 2005, dozens of countries met in Cambridge, England, to set up the Global Earth Observation System of Systems (GEOSS), to coordinate national systems and satellite observations into a single, global, earth-monitoring organization. Sixty countries participated, including all the major polluting countries.

Another curious indicator was the emergence of coal from polluting villainhood. In the year 2000, only two new coal-fired generating plants were planned in the United States; by the year 2004, there were no fewer than a hundred on order. Partly this was because the U.S. administration in 2000 and 2004 was resolutely non-Green, and partly because coal resolutely did not come from the Middle East, but that wasn't all. The technology had changed. Coal producers, stung by their unwashed reputation, have invested large sums in scrubbing technologies, and many of them actually work. More-modern combustion techniques not only clean the emissions before they start, but they burn less coal too. "A century ago coal plants delivered only 5 percent of the fuel's potential energy; now the number is about 35 percent, and pulverizing it can get that up to 40-45 percent. With high-temperature burns, over 50 percent may be possible." Joint industry-government research efforts in Australia and Canada had come up with a number of innovative ideas. For example, coal can be "fluidized" before combustion—you can burn it on a bed of particles suspended in air, a technique that captures most of the emissions before they begin. Coal can also be burned in oxygen and not in air; it can be gasified, with the gas powering a turbine, the surplus heat used to drive a conventional turbine. Noxious emissions can thereby be greatly reduced, perhaps to zero. As the *Economist* pointed out, much depended on how national legislation was framed. The Netherlands subsidizes zero-emission electricity; and Norway heavily taxes carbon emissions; both policies encourage the development of clean coal. But British subsidies, for example, are awarded only to renewable-source electricity, which leaves out even the cleanest coal burning.[36]

Finally, consider the question of the ozone hole over Antarctica, which only a few years ago was a serious cause for concern—legitimately, because while ozone (O_3, or oxygen with three atoms) is poisonous to humans at ground level, as it is a major component of smog, at high altitudes it protects the planet from the sun's damaging ultraviolet radiation. With ozone thinning so dramatically, the risks of rampant cancers and plant crop failures seemed very real; the ozone layer around the earth is thinnest at the tropics and thickest at the poles.

The hole in the ozone layer appeared over Antarctica very quickly. Or rather, by the time it was noticed in the early 1980s, the ozone layer had deteriorated so badly that the scientists who found the hole actually thought their instruments must be at fault and sent back to Britain, their home base, for a replacement set.

Ozone depletion was without question human-caused; the chemicals that were destroying it were man-made, mostly containing chlorine and bromine such as chlorofluorocarbon (CFC) and halogen compounds, none of which occur naturally. Natural sulfurous emissions from volcanoes also had an effect—but only by combining with the industrial chemicals already in the air to form chemically active clouds that dangerously accelerated the ozone-depletion process.

But in 1987 a group of industrial countries led, perhaps ironically, by the United States, met in Montreal and signed the Montreal
Protocol. This is one of the international community's greater suc-
cesses: not only is the manufacture of CFCs being phased out (they
are banned in most Western countries) but also the hole in the
ozone layer is recovering. There was a blip in the story when it was
discovered that some of the substitute chemicals—(hydrochloro-
fluorocarbons (HCFCs), hydrofluorocarbons (HFCs), and perfluorocarbons (PFCs)—were themselves powerful greenhouse gases, but
it was a blip that was swiftly overcome by further substitutions.[37]
Among the lessons learned: The rich world caused the problem and
must pay for the cleanup. And, secondly, that the poorer world must
agree to cooperate (or at least not to make things worse) but is right
to insist on both time and access to technology to help it adjust.[38]

Progress can be made, then.

Whatever we do to "fix" the problems we have caused, Ivan and
his grim successors will still be conjured into being by Aristotle's
"exhalations from the earth" and by the sun. Nor should we try to
prevent that from happening, because success would have unforeseen consequences for the planetary climate. But at least it's possible
for us not to make things worse. And we can learn to use what is
there, what *is*, to our own ecological advantage. All we need is the
wisdom to know what is advantageous and what isn't.

CHAPTER EIGHT

The Technology of Wind

*I*van's story: In the end, Ivan's eyewall just barely brushed the Cuban mainland, most of its central circulation staying offshore. It had taken yet another unexpected westward jog before turning north. The high-pressure ridge was proving more persistent than forecast, its weakness not as apparent. Nevertheless, with hurricane winds extending outwards 90 miles from the eye, and tropical storm winds 175 miles from the core, Ivan was a large storm and Cuba did not escape entirely. Pictures taken on a helicopter flight over the region a day later showed extensive damage to the popular scuba diving resort of Maria la Gorda, a favorite winter-time destination for paleskins from Canada, where several buildings had lost roofs and palm trees had been uprooted and scattered. Elsewhere, there was flooding and a small bridge and some roads were under water. There were no deaths or injuries to add to Ivan's fatality count, so far, of 68 (39 in Grenada, 18 in Jamaica, 4 in the Dominican Republic, 3 in Venezuela, 2 in the Cayman Islands, 1 in Tobago, 1 in Barbados). That this was a triumph of communist planning the tame Cuban media, for their part, had no doubts, though the news reports did mention both luck and the high-pressure ridge over which they professed no controls.

Ivan threaded its way through the Yucatán channel, and roared into the Gulf.

From a human perspective, this was about as benign a course as could have been hoped. A jog either way would have produced a death toll much higher.

The National Hurricane Center bulletin issued on Monday, September 13, predicted "close to 20 knots of westerly shear affecting Ivan but relaxes this shear a bit at 24 hours and 36 hours before increasing it significantly." As a result, the intensity forecast showed a gradual weakening.

The track forecast was still for a turn northward, and after seventy-two hours turning due north or even northeast as Ivan approached the prevailing westerlies, which had regained their stability as Frances expired in the Atlantic. That would still take it through the western fringes of the Florida Panhandle, through northern Georgia, and into North Carolina, though by then it was fully expected to have subsided to a tropical storm and then back into a mere depression—lots of water, not much wind damage.

The big question was—where would it hit the U.S. mainland first?

The Florida and Alabama emergency measures organizations wearily cranked up their evacuation procedures once again. There was the usual run on bottled water, portable generators, and sheets of three-quarter-inch plywood.

I had friends in New Orleans who were booking hotel rooms in Houston, just in case. New Orleans is a few feet below sea level, protected only by the fragile levees built to contain the Mississippi. The city hadn't had a direct hit from a Category 4 hurricane in living memory, never mind a Category 5—not, that is, until Katrina in 2005, which was a strong Category 4.

The technology exists to construct buildings capable of withstanding such hurricanes. But how do you rebuild—retrofit—New Orleans?

The Canadian Hurricane Centre in Dartmouth had been tracking the storm but had seen no need to issue public bulletins, either as warning or as reassurance. They were keeping on eye on Ivan, though. Hurricanes had done strange things before, and no doubt would again.

At five A.M. Eastern time on Tuesday September 14, Ivan had sustained winds of 160 miles an hour, once again making it a Category 5, but a Hurricane Hunters reconnaissance plane that penetrated the eye an hour earlier had measured a pressure of 924 millibars, slightly higher than before. The pi-

lots reported that the eye was well defined with very cold convective tops, but Ivan was nevertheless expected to weaken before hitting the coast. The track forecast, hedged about with cautions as it was, nevertheless showed the landfall now missing Florida and coming ashore on the tiny stretch of Gulf coast owned by Alabama. New Orleans was still watchful—and my friends had indeed locked their apartments and decamped for Houston, laptops in hand. The resorts along the Mississippi coast were shutting down as their customers fled.

By the following morning, with Ivan still twenty-four hours from landfall on the U.S. mainland, the National Hurricane Center was issuing bulletins every few hours. At five A.M. the wind strength was 140 miles an hour, making it a Category 4, but nearing the category's lower threshold. All the models now agreed on the track—it was going to hit Alabama overnight or on Thursday morning. Satellite images showed the center was weakening, and pressure was up to 935 millibars.

A new note of caution was introduced at this time—once Ivan crossed over the coast, its steering currents were expected to collapse, and the storm could either stall entirely or wander erratically to the southern Appalachians, giving up its huge amounts of Caribbean and Gulf water in torrential downpours, with the consequent risk of flooding. The warning arc was extended inland—because of the storm's size and intensity, it was likely to still be a hurricane up to twelve hours after landfall, about 120 miles inland. That possibility—and the erratic wandering expected—caught the attention of weather people all along the eastern seaboard, all the way up to Dartmouth, where their scrutiny of the storm intensified.

In the hours before its landfall on the U.S. mainland, Ivan was once again penetrated by pilots from the Hurricane Hunters squadron. They reported that the southwestern quadrant of the eyewall had all but disappeared. Ivan was finally losing its potency.

But it was too late for the Alabama coast and its barrier islands. And it was too late for the Florida Panhandle, which once again, for the third time in a month, was battered by high winds and torrential rain.

Ivan was still generating winds of 130 miles an hour as it crossed the coast at two A.M. That made it a strong Category 3, just under the Category 4 threshold.

The northeastern Gulf landscape is flat; the communities of Orange Beach and Gulf Shores, which were in the center of the eye as it came ashore, are both low-lying, their protective dunes only a few yards high, much too small to fend off Ivan's twenty-foot storm surge. The sea either brushed the dunes aside and crashed into the houses and marinas beyond, battering them away, or pushed whole dunes into the communities, swamping them in a swirling miasma of sand and salt water. The Interstate 10 bridge near Pensacola collapsed at the height of the storm; dozens of small boats secured in their marinas were dashed half a mile inland; the Alabama Gulf Coast Zoo was destroyed, and among the animals flooded out of their homes was Chucky, a one-thousand-pound alligator, and eight of his friends. Two suburban swimming pools were found floating on the highway near Gulf Shores.

By the morning of September 16, 26 more people had died, 15 of them in Florida. More than a million people were without power in eight states. (Later it was estimated that another 31 deaths were "indirectly" attributable to Ivan, including 6 in Canada, bringing the monster's death count to 125, and Ivan had spawned no fewer than 111 tornadoes across five states, destroying thousands of homes.)

By the afternoon of the 17th, the last workday before the weekend, Ivan had—finally—dropped below hurricane strength and had been reclassified as a tropical storm, and then downgraded further to a tropical depression. But it continued to spin off tornadoes and thundercells, not so very different from those that were at its core almost three weeks earlier, all those thousands of miles away in the Sahara. Nine inches of rain dropped on Georgia, causing widespread flooding. In middle Tennessee, several communities were hit by winds of almost hurricane strength, downing trees and power lines. Highways near Lawrenceburg were closed. In West Virginia, more than twenty-two counties qualified for federal disaster relief funds. Eight inches of rain fell across central Pennsylvania, and those residents who had managed to sleep found when they woke up that radio and TV stations were off the air, and hundreds of homes and cars were underwater. The village of Spring Mills was two or three feet under water. Dozens of rivers were between four and six feet above normal. (Many of the news reports, curiously, seemed more concerned that Schnitzel's Tavern in Bellefonte was under water. To outsiders this seemed a low-grade emergency, but no doubt Pennsylvanians believed differently.) Most of the Delaware River basin got half a dozen inches of rain in a few hours; the river and its tributaries, especially in the Catskill and Poconos mountains, swelled and overflowed. The main-stem Delaware river at Trenton, New Jersey, was at 298 percent of normal, the highest levels since the state got hit by back-to-back hurricanes in 1955. Several basin counties in Pennsylvania, New Jersey, and New York were declared federal disaster areas, and also qualified for relief funds.

I

Over their long history, humans have learned to use the wind in two primary ways, for transportation (sailing) and for supplementing their own musculature (that is, for driving machinery), and have learned from the wind (or other creatures that use the wind) another by-now-indispensable technique: the art of flying.

But sailing has become a game, in modern times a trick for children or a diversion for wealthy adults. As for flying, we are in the air more than ever—but as much against or despite the wind as with it or for it. And for using the power of the wind? Windmills have come, have gone, and are once again beginning to fill our landscapes, albeit in different iterations and guises.

Many other creatures, with a history far longer than ours, have also learned to use the wind, often in astonishingly subtle and complex, if rather limited, ways. In the course of doing so, they utilized, in passing as it were, some of our most cherished technologies long before we did—indeed, long before we existed. Not just flight, but navigation devices, echolocation, the barometer, the technique of sailing, the playful games of parasailing, parachuting, and gliding. They invented techniques for extracting scarce moisture from the air: In Namibia's Skeleton Coast, one of the most arid places on earth, a fog-collecting beetle uses the wind to condense moisture into little runnels on its wings, which it then funnels into its mouth. Termites "invented" air-conditioning. Some mammals have learned to imitate the insects—the Saharan jerboa, a burrowing animal, uses ventilation channels to move air through the warren and cleanse it.

There is more to life in the lower atmosphere than birds. Much, much more life, a riotous carnival crowd of life . . . The air in the first thousand yards or so off the ground is filled with a dense crowd of wind-blown pollens and fungal spores, as well as myriad insects and the birds that devour them; at some estimates, the air above grasslands in temperate zones can carry almost a million insects, or at least organisms of one kind or another, per square mile at any one time.[1]

The earliest creatures to learn to use the wind were the plants. Spores and fungi, which are ancient even by plant standards, depend on the winds for their movements and their propagation. All orchids, which are really just host colonies to fungi, still use wind to propagate; in some species a single flower can produce four million seeds, capable of surviving wind-borne trips of up to 1,800 miles. Other stay-at-home blooms, including some orchids, collect their nutrients from afar: I've already remarked on a species of orchid growing in the rain forest canopy of the Amazon that depends on African dust for a large share of its nutrients. And allergenic plants such as dandelion, ragweed, and goldenrod cause outbursts of sniffling in the spring, as their pollen is carried on the winds.

Many large plants use the winds to disperse seeds. Canada's national tree, the maple, spreads its seeds on little winged husks, which will carry long distances in a small breeze. Conifers spread their pollen in the winds. Whole plants, too, use the winds. The tumbleweeds of arid areas are entirely wind-dependent for their travel; one of our coaches at school used to make us chase tumbleweeds across the dusty plains in the Orange Free State, the better to get us fit for the hockey season (well, it beat doing laps).

Winds can bring alien invasions as well as benefits—see those soy-eating fungi invading Arkansas from South America. In Cape Town, one of the windiest places on the planet, agronomists imported the Port Jackson willow from Australia in a vain attempt to anchor the sands of the so-called Cape Flats, which were threatening to blow away and turn Cape Town proper into an island. It spread rapidly in the winds, and is now a threat to native species; my sister is one of thousands of Capetonians who turn out in organized "hacks" to root out this and other aliens. As Jan DeBlieu points out, Hurricane Andrew in 1993 knocked down so many native trees in Florida it helped the spread of four alien imports: melaleuca, Australian pine, Brazilian pepper, and marlberry.[2]

Insects, too, have worked out dozens of ways of using the winds. Some of them do it by simply getting aloft and being blown along. In 2004, locusts invaded Egypt again, just as they did in biblical times, using the prevailing winds to do so. Many otherwise wingless insects do the same. At sea, certain microorganisms are thrown into the air by wave action, and are carried enormous distances on the winds in aerosols.

Other insects have learned to use wind in more indirect ways. That spiders can employ wind to cross fairly long distances between trees, dangling themselves on the end of a long rope of silk, is commonplace, and can be seen in pretty well any backyard. But entomologists have also identified some species of spiders that can build simple sails from their silk to lift themselves into the air. To some degree they can control their flight by lengthening or shortening the threads they produce.[3]

II

One autumn morning, after yet another gale had passed, I sat with a coffee in the atrium of our house watching the gulls circling high overhead. Angry clouds were still scudding by and the winds were still strong, but the storm center was safely to our northeast and the mood was more relaxed. The gulls, it seemed to me, were also relaxed, playful after the storm.

This is a bit of a thorny issue, this business of a bird's playfulness. Ornithologists deride the notion and fishermen, who are used to gulls flocking about when they are cutting bait, say the birds are only looking for food, but I don't believe them. As I watched, five or six gulls would ride the wind, deftly matching the lift of their wings to the strength of the gusts, and were able to remain absolutely stationary in the air over the beach, sometimes for minutes at a time. Then one or another would peel off, dive downwind, and come back up again, to resume its place with the others, motionless in the wind. Never once did I see them dive down for food, or even seem to be looking for something. Perhaps there is some obscure Darwinian purpose to all this. Perhaps they are merely airing out their feathers. Perhaps they are proving to potential mates what terrific fliers they are, but it still looks like playing to me. Why not? If natural selection has given them these superb wings, what is so outré about the notion that they are actually enjoying that with which they are blessed? Indeed, the ornithological orthodoxy seems to me unnecessarily dour.

In the summer you can see the ravens playing in the winds too. A raven will balance, motionless, on an updraft, until, with a subtle change of windspeed or just a decision made inside its dark skull, it will shift its wings to ride an invisible crest of air at great speed. As British nature writer Paul Evans puts it, "the most dramatic displays are when a raven launches through the wind, rolls, flips over and back, then fans out the wings and tail in a mighty swish to soar away with great insouciance."[4] Near our house the ravens soar, clasp talons with a fellow, and then . . . tumble, in a raven game of chicken, to see which bird will unclutch the other first before risking being dashed against the earth. Again, I suppose it is plausible that they are merely demonstrating bravado in a kind of dominance ritual dictated by natural selection and the need to impress girl ravens and therefore sire young, but perhaps this (imagining them as grotesquely competitive as humans) is the anthropomorphizing, and not the playfulness.

Whatever the truth of the matter, it is certainly a fact that birds—and insects too—have learned over the long millennia of evolution to live in the wind in a way that humans cannot, to live in it, understand it, and use it, as transportation, a source of food, as a locus for sexual adventure, and even for foretelling, since many birds seem able to predict the weather from the winds. Indeed, some birds spend most of their lives in the air; albatrosses, for example, seldom land; a new study in 2004 found that albatrosses routinely circumnavigate the globe twice in a season, even sleeping in the air, maintaining themselves aloft with some sort of natural autopilot. They only come down to feed.

Still, none of these astonishments has attracted the human imagination more than the apparently effortless way birds take to the air—*pace* our thoughts on those gulls hanging motionless in the breezes on the shore, or the way the great raptors use updraft thermals (anabatic wind systems) to gain altitude without expending any energy.

It looks so *easy*.

The gulls I had been watching disappeared by late morning. Not necessarily because they were tired of the game, but a couple of them had spotted one of our lobsterman neighbors rebaiting out on the bay, and were circling overhead waiting to see what could be gained. Typically, for gulls have superb eyesight, within minutes a dozen or more appeared, flapping vigorously upwind, from where they had been sitting on the rocks out on Coffin Island. They took up station over the boat, like patrol aircraft on security duty, cruising in lazily effortless circles.

Flapping, circling, cruising, diving, landing, all part of their native technique, all come to them hardwired into the genes. The naïve view of how birds fly, the common-sense view that still feels right to me, in some stubborn corner of my mind that is resistant to apparently overcomplicated science, is simply that the flapping of their wings somehow pushes the air down, and therefore them up. I knew that aircraft didn't flap their wings, though not for lack of trying among early inventors of ornithopters and other curious vehicles, but I just figured they worked in the same way that your hand is pushed upward when you hang it out a car window at speed and angle it just slightly into the wind. The wind has force, we know that from . . . well, from wind, the force of wind on trees and other objects.

The reality is more complex. If you examine a bird's wing in cross-section, it is generally flattish on the bottom and curved on the upper surface; the curvature is more pronounced in some birds than others, but it is always there, at least among those birds who still use their wings for flying, unlike ostriches, or the late-lamented dodo. That curvature, it turns out, is critical. The reason for it was worked out as long ago as the eighteenth century by Dutch-born mathematical physicist Daniel Bernoulli, although he didn't apply it to flight. He was concerned with more prosaic matters, like water pressure.

Bernoulli and Blaise Pascal pretty well invented the science of fluid mechanics—and hence the study of laminar and turbulent flow, and hence aerodynamics. Bernoulli's principle, which sounds deceptively simple, was that the pressure exerted by a moving fluid (water, or air) is a function of the speed at which the fluid is moving. This is

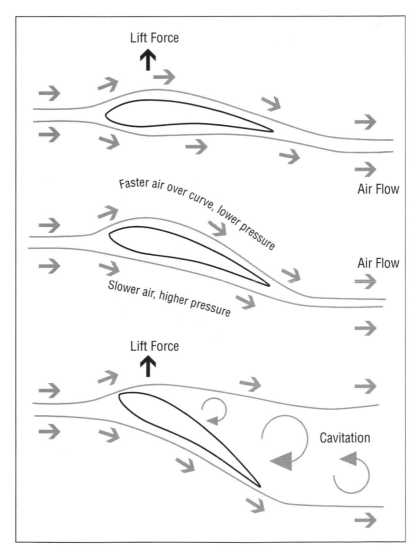

Bernoulli's principle, showing how air flowing faster over a curved surface creates lower pressure, and therefore lift. When the angle of attack is severe, cavitation is caused behind the upper surface, which exaggerates the lift force and can cause a wing to "flip."

the familiar garden hose effect that was noted in the discussion on wind force; the pressure of the same volume of water through a hose will vary depending on the diameter of the pipe—the nozzle, if you like. So if air is flowing faster on one side of an object than on the other, the faster flow will cause reduced pressure in that area, and consequently the slower side (high pressure) will be pushed toward the faster side (low pressure). Bernoulli's principle explains a good many humble phenomena. For example, it explains why a shower curtain is sucked in against the showerer—the flow of the water from the shower head decreases the pressure inside the curtain. A curveball in baseball depends on Bernoulli's principle too. The spin imparted by the pitcher causes air to move more rapidly on one side of the ball than on the other, increasing the pressure on one side, with the net effect of pushing the ball off course. Once aloft, birds obtain lift in exactly the same way. Air flows over the curved upper surface of the wing more rapidly than it does over the flatter lower surface; as per Bernoulli, the more rapid flow of air above the wing results in decreased pressure there, allowing the normal pressure beneath the wind to push upward. Birds, like planes, fly by being pushed upward by air pressure.[5] They don't lift, they are pushed.

Bird wings are rather more complicated than simple aircraft wings, which is not so surprising, given they have developed over many millennia of trial-and-error flying. Also not surprisingly—since birds probably developed from flying dinosaurs, and flying dinosaurs developed from quadrupeds—wings are really vestigial arms. Or, looked at from an avian point of view, not so much vestigial as more highly developed arms. As a consequence, bird wings, unlike those of insects or of aircraft, consist of two quite separate parts: an arm wing and a hand wing.

It is the arm wing, the part closest to the body, that yields up the conventional aerodynamic profile—a rounded leading edge, and a curved upper surface. In birds that fly long distances like albatross, the arm wing tends to dominate because the profile provides high lift and very little drag at fairly high speeds. But in small, agile birds, the hand wing plays the dominant role. And no birds are more agile in the air than the nimble swifts, little birds with swept-back wings that hunt insects and catch them in flight. Of all birds, swifts can turn on a dime, brake more sharply without stalling, and accelerate more quickly. How do they do it?

By contrast with its arm wing, the leading edge of a swift's hand wing is sharp, nothing more than the narrow vane of the outermost primary feathers. A group of Dutch biologists decided to check out how the hand wing works, using what they called "digital particle image velocimetry," not in a wind tunnel but in a water tunnel. Apparently the swifts wouldn't cooperate by flitting about a conventional wind tunnel in a straight line, so the Dutch scientists built an artificial swift wing instead. The results surprised them.

Earlier studies of insect wings had discovered that some creatures, masters as they were of the unconventional lift (remember the old saw about it being physically impossible for a bumblebee to fly?) generated what were called leading-edge vortexes, which greatly exaggerated the upward push of the flowing air during both flapping and gliding. A leading-edge vortex forms when the so-called angle of attack, the angle between the wing and the incoming air, is fairly large. The air flow then separates from the wing at the leading edge, and rolls up into a vortex. To facilitate vortex formation at lower angles of attack, the wing needs a sharp edge. To exploit the vortex, to use it to get lift, the insect or bird must keep this rolled-up vortex close to the wing. Insects typically do this by very rapid wing movement. Swifts do it by sweeping back the angle of the hand wings, almost to a V shape. As a consequence, the leading-edge vortex, or LEV, spirals out toward the tip of the wing, looking for all the world like a tiny tornado. And as with real tornadoes, the air pressure at the core of the vortex is very low, sucking the air beneath the wing upward, giving extraordinary lift. These vortexes are also remarkably stable at both low and high angles of attack—insects generally need an angle of about 25 to 40 degrees, but swifts can operate successfully at angles as low as 5 degrees. This gives them the speed to accompany their agility, and the ability to catch quickly flitting insects. Swifts change the sweep angle of their wings in flight, thus changing the angle of attack of the air flow. They use low angles for speeds, high angles to brake in midair—it gives them lots of drag, but the vortex keeps them from stalling and losing height. Aerospace engineers have copied the principle for certain military aircraft, which must be highly maneuverable and perform well at both subsonic and supersonic speeds; pilots of the newer fighter jets, such as the Tornado, can choose different sweep angles for agility or for cruising speed.

The next challenge, the Dutch scientists say, is to learn more about how swifts use their variable wing sweep to directly control the leading-edge vortexes to increase their flight performance. "The swift's flight control might inspire a new generation of engineers to develop morphing microrobotic vehicles that can fly with the agility, efficiency, and short take-off and landing capabilities of insects and birds."[6]

O f course, this isn't the first time humans have learned from birds—or, in the early days, attempted to learn from birds. Icarus is a notorious example, though he is now remembered mostly for his hubris rather than his flight-control technology. Dreamers from Roger Bacon through Leonardo da Vinci have sketched vehicles that somehow mimicked birds, some of which would no doubt have worked if the engineering skills to make them had existed and if the knotty problem of takeoff had been solved. Takeoff was always the most difficult issue. The early inventors solved it by the simple but rather hazardous expedient of hurling the winged one over a cliff.

Leonardo's sketchbooks show many gliding devices, most of them winglike objects strapped to a human body, looking curiously like modern hang gliders. In this as in many other things, Leonardo was precocious, and not much progress was made for another four hundred years or so. In the early nineteenth century George Cayley designed and built the first true glider, a small biplane made of cloth sails with a horizontal tail and two lateral fins. One of his designs carried a man about 900 feet.[7] All through the nineteenth century the occasional dreamer attached wings to his body and leapt off a building, and occasionally survived; but hang gliding as a sport had to wait until the twentieth century, when NASA engineer Francis Rogallo and his wife built a wind tunnel in their home to develop a personal flying device consisting of a delta-winged sail controlled by ropes. Since then, activities using personal wind-powered devices have proliferated. On the ground, they include soft-winged sailboards and rigid-winged kitewings (which can reach 24 miles an hour in good conditions), kite surfing, skate sailing, and ice sailing. In the air, parasailing, hang gliding, and wingsailing.

The Chinese were early experimenters with flying techniques. That they invented kites a long time ago, probably before the fifth century B.C., we know because philosopher Mo Zi, who lived between 478 and 392 B.C., made a wooden kite in the shape of a hawk that flew for a whole day. Some reports, dating back much earlier than that, indicate that the Chinese used umbrella-like devices to jump off towers or high mounds. But the first modern parachute descent was in 1783, by French physicist Louis-Sebastien Lenormand, who hurled himself off the tower of Montpellier Observatory and drifted safely to ground. Two years later Jean Pierre Blanchard ascended on high in his balloon, attached a parachute to his dog, and dropped it from several hundred yards. The dog landed safely but was said to have run away and was never seen again.

It was the gliding flight of storks, slow and stately birds, that inspired the first aircraft designs of Otto Lilienthal in the late nineteenth

A flight of Otto Lilienthal

century. Lilienthal, who was one of the inspirations for the Wright brothers, built what he called a sailing apparatus very like the outspread pinions of a soaring bird. "It consists," his notes say, "of a wooden frame covered with shirting (cotton-twill). The frame is taken hold of by the hands, the arms resting between cushions, thus supporting the body. The legs remain free for running and jumping. The steering in the air is brought about by changing the center of gravity. This apparatus I had constructed with supporting surfaces of ten to twenty square meters. The larger sailing surfaces move in an incline of one to eight, so that one is enabled to fly eight times as far as the starting hill is high."

Lilienthal's reputation as a respectable scientist at last pushed experimentation with flight beyond the province of dreamers and fools. Over a span of five years he developed eighteen models of gliders, fifteen of them monoplanes and three biplanes. Each was essentially a hang glider, controlled by the pilot shifting his weight.

"To invent an airplane is nothing," he once famously said. "To build one is something. But to fly is everything." To facilitate his flights, he built a conical hill in his backyard at Lichterfelde, near Berlin, so he could launch his gliders into the wind no matter which direction it was coming from. Alas, he was a victim of his own experiments, and he died after a crash of one of his hang gliders on August 10, 1896.[8]

After the Wright brothers, as we know, aviation developed at an extraordinary pace. By 1908, maybe ten people in all the world had been in an airplane. Four years later, literally thousands had flown. At the beginning of the twenty-first century, there were some eight thousand commercial flights aloft at any one time somewhere on the planet, carrying maybe a million people.

We're still learning from birds too. U.S. engineers are planning an aircraft called the Pelican, which is based on the way that those slow-flying and often low-flying birds exploit what is called the ground effect—a curious phenomenon in which close proximity to the earth's surface actually reduces in-flight drag, and increases the upward efficiency of the wing. The Pelican aircraft would be huge—on the drawing board is a version four hundred feet long, with a five-hundred-foot wingspan and a cargo capacity of around ten Boeing 747s. It would fly at a stately 250 miles an hour at an altitude of no more than twenty feet—the first major commercial airplane to have to keep a close eye out for icebergs.[9]

III

A few days after the autumn 2004 storm had passed, we went down to the wharf in the little village of West Berlin to pick up some lobsters from one of our neighbors, Bob Lohnes. The whole fleet was in, all four boats—the West Berlin flotilla is not exactly a threat to global fish stocks. They were all pretty similar, Cape Islanders, a roofed cuddy amidships, an open deck for stacking the metal wire lobster pots, crates for the lobsters themselves, a winch for hauling the pots. The motors were sturdy diesels with a throaty sound; the old one-lungers of the East Coast, with their open spark you could light a cigarette from, have long since retired, nor do modern fishermen any longer shoehorn old Chrysler truck engines into their hulls for motive power. The Cape Islanders are small boats, but very sturdy, maneuverable, and stable even in rough seas, perfectly tuned for the job they are designed to do.

For our purposes, though, the most notable thing about them was the absence of sail. None of them carried any canvas except sometimes a tiny "staysail," used only to keep them into the wind while hauling traps. GPS systems, radars, range finders, echolocaters, cell phone chargers, yes, but no sails. There was no longer any point to sails.

I looked across Blueberry Bay. It was empty except for the sprightly colored buoys that marked where the lobster pots were. Farther out there was a coast guard vessel with a red-striped hull. Otherwise . . . nothing.

In summer it is different. A small marina has been built to the west, in the town of Liverpool, and on summer afternoons little triangular scraps of white bob perkily out from the Mersey River and head across the bay, perhaps bound for Lunenburg or Chester. Less frequently, and usually in the fall, slightly larger triangles appear to the east and head across the horizon, larger yachts, some of them heading for Bermuda and points south. Rich peoples' boats; a whole subculture of self-described "yacht bums" travels up and down the eastern seaboard, making a living of sorts crewing other people's boats to exotic ports. But none of this is any longer necessary. None of it is "commerce." Instead, it is all recreation. Many of the people who sail these little boats are very skilled, but their skill is employed for its own sake, in no cause and to no real purpose. In the old days sailing wasn't fun, something you did after work. Sailing *was* work. It was the lifeblood of world commerce, and wind was the oil of the time, the motor that drove the engine of global trades. And of course it was free, and lasted forever, the earth as a perpetual motion machine.

All gone now.

But the age of sail, which started the age of globalization, lasted for many millennia, much longer than the upstart machine age that has succeeded it.

Sailing probably began along the Nile, or possibly on the Euphrates delta of old Mesopotamia, or the Middle Kingdom of the Han. The oldest extant pictures of sailing boats are from the pharaonic cultures of Egypt, dating back some four thousand years. The civilized world's first known shipwreck is depicted on a stela at Karnak, dating from the second millennium B.C. Boats eerily like it are still visible plying the river only hours from Cairo. On modern Lake Nasser, formed by the Aswan High Dam, the feluccas ply the waters as they have done on the "Longest River" for forty centuries. Higher up the Nile, at Lake Tana in Ethiopia, they are called *tankwas*. They are made of papyrus, and except for their lateen sails are identical in style to the boats depicted on the frescoes at Luxor and Karnak. You can sit by the Nile at Aswan and if you are lucky, you may see the feluccas drifting by in the light of a blood red moon. The Pharaohs pushed their empire steadily south passing Abu Simbel and Dongola and the Second Cataract, as early as 2300 B.C., conquering the settled kingdoms they found there, and by 2000 B.C. Nubia was under Egyptian control. The gold trade increased and with the gold came wealth, and commerce on the river increased. By the second millennium B.C. boats were coming down from Meroë and Kush, as high as the Fourth Cataract, carrying trade goods, livestock, and piles of what look like lumber in the frescoes. They already carried square-rigged sails amidships.

By Homer's time, sails were ubiquitous in the Mediterranean world, and traders were fetching cargoes from Africa and the Levant. The most sophisticated sailors of early history were the Phoenicians;

it was their skills with sail that earned them their dominance of the western seas. Some evidence suggests that the Phoenicians made their way into the Red Sea and thence as far down the coast as Zanzibar. They certainly rounded the bulge of Morocco into the Atlantic, and by some reports made a circumnavigation of Africa a thousand years before Vasco da Gama left his markers on the shore in Table Bay, in what is now Cape Town.

Sailors used the prevailing winds, and the rivers of the sea, the ocean currents, to get around. They could only sail downwind, and either had to row back or to catch another wind somewhere else, and a further one to take them home. In the Mediterranean, a map of the prevailing winds soon appeared. Arab sailors used the monsoons to take them to India and the more southerly easterlies to bring them back. The Chinese used the currents and the prevailing winds to thread their way through the islands and isthmuses of southern Asia, and across the Indian Ocean to Sofala, then the Swahili-dominated portal to the ancient African empire of Zimbabwe. The lateen sail, a curious triangular thing on a movable boom, was developed somewhere in the Far East, made its way through the Middle East by way of Arab traders, and finally appeared on Roman ships a few decades before Christ. This made vessels much more maneuverable, and less dependent on the direction the winds blew. Even so, you couldn't sail into the wind, or even close to it. You wouldn't be able to do that for another thousand years at least.

The history of exploration is the history of sail and therefore of wind exploitation. The Norse knorr and its successors the cog and the carrack and the caravel made global exploration possible. We know the Norse reached Newfoundland by around A.D. 1000; Basque and Portuguese fishermen were not far behind. The Norse used the subpolar easterlies to head for America, and the midlatitude westerlies to get back, at least until the mini ice age, when the polar seas filled with ice, and transatlantic voyages perforce had to wait for

Columbus, who would use the tropical trade winds for his crossing. The Chinese developed similar vessels at about the same time, probably independently. The notion that the Chinese discovered America for outsiders long before Columbus did has been propounded (and not—yet—debunked by historians). If they did so, they would have used the clockwise Pacific gyre, both current and wind, to get there. It was certainly true that for a brief period, about 1300 to 1400, the Chinese mastered the oceans, building 350-foot vessels with nine masts; the Mings once raised a fleet of more than 3,500 ships, mostly for trade—they rather disdained conquest, not wanting too much truck with lesser cultures.[10] By the fifth century, Indian and Indo-Malayan merchant ships traveled from ports on India's east coast to Guangzhou (Canton); later, Arabs and Persians sailed from the Red Sea to India. By the eighth and ninth centuries China's southern ports were already full of foreigners; merchant ships that plied the China Sea carried both oars and sails, and already used the compass for navigation.

From the thirteenth century onward, vessels both Occidental and Oriental were capable of sailing anywhere in the world. The workhorse of the trading world was the caravel, a ship about seventy feet long and with three masts: the foremast carrying a square foresail and topsail, stepped through the high forecastle at the bow; a mainmast, amidships, with square mainsail and topsail; and aft, on the raised sterncastle, the mizzenmast, with a lateen-rigged sail. With the exception of the lateen sail, the sails were hung from a yard at right angles to the longitudinal axis of the boat. The lateen sail at the stern, set fore and aft, had several advantages over the square sails. It was more efficient in sailing close to the wind, and it could be used to push the boat around when tacking.[11] The increased number of sails also created a boat with greater maneuverability, one that was faster, easier to sail, and required fewer crew. It became the standard vessel for Prince Henry's discoverers and was used by Columbus in his explorations.

For the next few hundred years shipbuilders produced as many new designs as Detroit did cars in the 1950s. Sails and sail types proliferated. Before long, a typical ship's mast would carry six sails—the "course" at the bottom, the lower and upper topsails, the lower and upper topgallants, and the royal or skysail at the top. Vessel types proliferated too. There were barques, barquentines, galleons, East Indiamen, frigates, brigs and brigantines, snows, and then schooners.

> *A full-rigged ship is a royal queen,*
> *Way-hay for Boston town, oh!*
> *A lady at court is a barquentine,*
> *A barque is a gal with ringlets fair,*
> *A brig is the same with shorter hair,*
> *A topsail schooner's a racing mare,*
> *but a schooner she's a clown, Oh![12]*

The schooner was developed in the Boston states, which in practice seemed to include Nova Scotia, as the fishing and trading workhorse of coastal waters; its ancestor was the two-masted coaster that plied British and Dutch waters in the sixteenth century. In the nineteenth century, Lunenburg and Shelburne and Yarmouth, Portsmouth and Gloucester—especially Gloucester, Massachusetts—and Boston found themselves the center of one of the world's most productive industries, building more ships in a few decades than any other place but Old England, not just schooners but brigs and brigantines and barques. The schooners that emerged from shipyards up and down the coast generally had two or more masts, without any square sails. Most were small, nimble, weatherly craft, but some had seven masts and were among the largest sailing vessels ever built. The topsail schooner, a British development of the American schooner, did carry one or two square sails on the upper part of the foremast, which improved her downwind performance.

The age of sail ended with a panache sel-
dom exceeded in any technology. For one brief decade, in the late
1840s and early 1850s, the clipper ship burst her way through the
stodgy and the hidebound, and raced her wondrous way into men's
hearts. As John Dyson put it eloquently in *Spirit of Sail: On Board
the World's Greatest Sailing Ships*, "The Clipper was a ship to grapple
with every element but fire. In the whole history of navigation
nothing excelled her dash and good looks—the slender hull,
springy as a sea hollow; the three tall masts slightly raked to give
her a youthful look, hungry for action; the great blade of her bow,
curved and sharp, scattering flying fish as it scythed blue water."

Before the clipper, commercial long-distance sailing was still a
relatively ponderous, slow, methodical, and mundane affair. As
Dyson says, British maritime law still mandated that "British cargoes [must be] carried on British keels," a way of keeping the upstart Yankees out of world trade. The great merchantmen, like the
East Indiamen, were cumbersome, slow, heavily armed, more akin to
a warship than a merchant. "Their officers wore naval uniforms,
and their heavily armed gun decks were manned by naval gunners
to fight off the roving pirates of the Arabian and China coasts . . . It
took a gale of a wind to move one of these elephantine ships, and
when they did move, it was seldom at the rate of more than three to
four nautical miles an hour." What fast vessels there were, sloops,
cutters, and schooners, were little bigger than modern yachts, almost
always less than one hundred feet long.

"But suddenly a new ship appeared, the Yankee clipper. She was
long and lean, with a beautiful, sweeping sheer line, and such clouds
of snowy canvas flying from her lofty spars as to make the old salts
shake their heads and predict the clippers would capsize at their
piers before even getting under way."[13] The first clippers were built
in New York by an American consortium, the shipwright a young

man named Donald McKay from Jordan River, Nova Scotia, who as far as we know had only built one vessel before, a barkentine (and whose accounts with a Shelburne blacksmith still survive—he bought rivets for crosstrees, for mast hoops, and for rudder bands; hinges for quarter boards; a strap for a martingale; hoops for a windlass). Alas, he went bankrupt, and decamped like so many before and after him for America.

Then the British repealed the law demanding that only British-built vessels be used for trade; and this allowed the Yankees to sail triumphantly into history.

The Yankee clipper was not massive, and her cargo capacity was modest. Her genius lay in speed, and with her knifelike bow and sweeping lines, she cut through the water twice as fast as anything else afloat—faster even than modern steamers. Under the hard-driving captain Bully Forbes, the Boston-built *Lightning* logged 436 nautical miles in twenty-four hours in a southern gale. It is probably the fastest day's run ever made under sail.

The Yankee clippers, though, were generally made of softwood and soon got waterlogged and sluggish. The British, no sluggards themselves in the art of shipbuilding, took the idea and made their own Clippers from iron and hardwood, and the China clippers that resulted are regarded as the apex of the shipbuilder's art—the perfect combination of grace and beauty on the one hand and cargo-carrying and seaworthiness on the other. Perhaps the greatest of all was the *Taeping*, made of iron and greenheart oak and teak, which carried nearly 30,000 square feet of sail, and covered 16,000 sea miles in a paltry ninety-five days.

Both to and from Australia, the traditional route for sailing ships was by way of Cape of Good Hope; the duration of passage was typically 120 days, but clipper ships cut that in half. Outward bound to Australia they turned well south of the Cape and headed across the bottom of the world to run their easting in the powerful winds and wild seas of the south latitudes known as the roaring forties.

Running for home, they continued around the globe in the same direction, looking for strong stern winds in the world's loneliest ocean, then turned Cape Horn and headed up the Atlantic . . . The clipper ship's trade was distance. She was not capacious, so all her profit lay in speed. The breathtaking nerve and splendor caught the popular imagination as space flights do today; more bets were placed on her finish up the Thames than were placed on the Derby.

In the heyday of the China Clippers, it became de rigeur among the chattering classes in London to be the first to drink the freshest tea from China, debarked from the first vessel of the season to arrive—the Beaujolais nouveau of its time. In 1854 the *Stornoway* and the *Chrysolite* headed from Whampoa, China, to Liverpool, England, at a dead run, both arriving in exactly 105 days; the *Stornoway*'s skipper remained on deck the whole time, sleeping what sleep he managed in a chair lashed to a hatchway. The *Oriental* made it the following year in 97 days.

It couldn't last. The Suez Canal, "that dirty ditch," marked the end to the days of sail, and the black smoke of the steamers finally overtook the billowing canvas of the clippers. Most of the great ships met ignominious ends. "*Chrysolite* was wrecked in Madagascar with a cargo of bullocks; *Stornoway* foundered in the North Sea; *Staghound* burned to the waterline; *Surprise* was sunk by a drunken Japanese pilot; *Fiery Cross*, *Taeping* and *Serica* were wrecked in the China Sea; *Ariel* was pooped and lost in the southern ocean. Others were cut down by steamers, converted into coal hulks, or simply disappeared with all hands. Only the *Cutty Sark* remains, a dry docked hulk in London."[14]

This kind of loss is still poignantly felt around here. Among the many sleek and beautiful vessels turned out of Lunenburg shipyards was the legendary *Bluenose*, under its even more legendary skipper, Angus Walters. For decades the *Bluenose* raced against the best and fastest that New Englanders from Gloucester and Boston could throw against her, and though she lost a few races, even the Gloucestermen, albeit grudgingly, called her Queen of the Atlantic. But in the end, the internal combustion engine was not resistible. They tried packing a motor into the *Bluenose*, but she wasn't meant for diesel, and was sluggish underway. Eventually she foundered on a reef off Haiti with a cargo of coal, and was lost. That was in 1946. That really was the end.

Unless, unless . . . Don Barr, the former skipper of the tall ship *Bluenose II*, believes that anyone with a fifty-foot schooner could make a good living today, in the early years of the millennium, the cost of freight—the cost of oil, he means—being what it is. Never mind what the burning of oil is doing to the environment; its uncertain supply and spiking cost, he believes, may make sail once more competitive, not just for feel-good do-gooders, but for businessmen trying to cut their costs. And the experimentation goes on, for humans really can't resist tinkering to make things just that little bit better . . . and the wind is always a challenge.

In September 2004 a curious flotilla gathered off Rhode Island. The boats were all what is called C-class catamarans. They were all oddly shaped, with protrusions and struts and ailerons and peculiar sails that only occasionally looked functional. This was the International Catamaran Challenge Trophy, also known as the Little America's Cup, the world's most high-tech regatta. What attracted the high-techies was that the regatta's relaxed rules meant the designers could try almost anything as long as they didn't exceed the maximum dimensions and permitted sail area. It was the first time the race had been run since 1996, when Duncan MacLane of the United States skippered *Cogito* to victory over the Australians. Going in, the favorite was a British vessel called *Invictus*; designed by aerospace engineers and sailed by John Downey, a retired Concorde pilot. In trial runs *Invictus* had reached the astounding speed of 30 knots (34 miles an hour) in a 15-knot wind—the real America's

Cup yachts would be lucky to reach 10 knots under the same conditions. In the end, though, *Invictus* had to scratch after an accident in the setup races, and *Cogito* won again, beating its boathouse-mate *Patient Lady* in the finals.

The most interesting things about both *Invictus* and *Cogito* were their sails—a rigid wing that looked as though a real aircraft wing had somehow been chopped off and stuck upright on the hull. The wing sail works the same way an aircraft wing does, using lift provided through Bernoulli's principle, except that the lift is forward and not up, just as it is in a pitcher's curveball. This forward lift drives the vessel forward because the sail is held upright at an angle to the wind. The rigid sail is more efficient than canvas partly because it doesn't have to waste time finding its shape before propulsion happens, and the whole sail can be at the correct angle, not just the core of it. Also, its carefully designed shape produces lift at a much smaller angle to the wind than a fabric one. And finally, a rigid sail supports itself, which means less cabling and tension wires, which means everything can be lighter, like a sailboard. The vessel itself weighs little more than the two people that crew it.[15] The downside: You can't reef the sail in a gale. In a real blow, you stay home.

Apart from the invention of sailing ships and windmills, and then aircraft, humans were slow both to understand and then to use local winds, and slower still to copy natural examples readily at hand.

Take the example of air-conditioning—the cooling of uncomfortably overheated air. I've already mentioned that termites "invented" or discovered or at least used air-conditioning. When I was a boy, we once demolished a termite mound (not out of malice, only practicality—crushed and rolled out, termite earth makes the best "clay" tennis courts in the world, and my uncle Blen was paying us pennies a wheelbarrow-load), and I saw for myself how the insects had angled ventilation chambers into the wind to bring cooling air deep down into the earth. They had even invented pressurization; there were dead-end chambers where the winds were compressed before being redirected even deeper in the mound, deep down where the queen lives. Termites invented air-conditioning, what, a hundred million years ago? Humans had to wait until fairly recent historical times for a version of it. Early human-created air-conditioning systems simply consisted of hanging damp rags in windows and doors, where air currents would cause evaporation, and thus cooling, an effect arrived at empirically, with no knowledge of the mechanics involved. Later, the system was reinvigorated by the Roman emperor Varius Avitus, who ordered ice and snow from nearby mountains to be placed in public parks for the same reason.

Other cultures have developed other devices for cooling the air, or at least fending off the extreme heat. The desert nomads in the Sahara, where damp rags are not an option, developed a simple wind flow device consisting of a horizontal layer of fabric suspended on poles above a tent, which has the effect of creating differential heating patterns, which produce a breeze between the two layers, muting the brutal heat of the Saharan sun. Air cooling dictated the layout of the Egyptian city of Kahun in pharaonic times; at around 2000 B.C. the Kahunian power elite made sure their houses were oriented to the cooler north winds, while the slave classes were packed in higgledy-piggledy to the south. In pre-Raj times, the city of Hyderabad in India contained houses with tall central air shafts and air scoops on the roof oriented to the winds, that drew cooling air into the interior. This was the same pattern developed, or imported, by the Swahili traders of Zanzibar, a system still used in that city, where the stone houses tend to be five or six stories tall, with the cooler sleeping rooms on the lower levels and the warmer public rooms higher up (the kitchen is typically on the roof). The Romans used similar ducts for heating, as the Incas did for their smelting furnaces.

In this sporadic, episodic way, a technical mastery of the winds developed. Human cultures moved quickly beyond having to sacrifice virgins to placate the wind gods; even Aristotle's sketchy knowledge of meteorology represented real progress, in the sense that it sought a technical grasp of how wind actually worked. But for an understanding of the theory behind it natural scientists had to wait until Leonardo had grasped the principles of conservation of mass; and even then nothing could be confirmed until Torricelli, Galileo, Sir Francis Bacon's *Historia Ventorum*, and Isaac Newton's theories of mechanics. As late as the nineteenth century engineers were still operating with hazy theoretical principles and had to resort to actual testing to see what was needed. A good case in point was Gustave Eiffel's design and construction of the Eiffel Tower for the French Exposition, which led to considerable advances in atmospheric science—Eiffel's wind-load design assumptions were among the earliest sophisticated attempts to understand static wind loading on buildings.

It took another five decades before the first wind tunnels were built for the laboratory testing of winds, and it wasn't until the 1970s that the term *wind engineering* took on common currency. But now, in the laboratories of the great universities, not just in the faculties of environmental studies but in the schools of engineering and applied sciences, the seductive notion that the wind is essentially free motive power is once again taking root. There is still resistance to the reality of our negative impact on the planet, but the oppressive weight of the evidence is having its effect and experimental notebooks are filling with curious designs that are at once sleekly modern and archaic in conception. The engineers playing with their catamarans in the Little America's Cup are but one example. Sail-assisted ocean steamers had a brief fad in the eighties, and are now reappearing; there are already models capable of reducing fuel burdens by up to 15 percent. More radically, self-fueling vessels are on the drawing boards, driven by hydrogen engines, the hydrogen derived from the sea via wind power. Dirigibles, no longer hydrogen filled but hydrogen propelled, are also reappearing. Aircraft designers are looking to thermal lift, as the birds always have.

The designers haven't gone back to Icarus yet, but it can't be long.

IV

Lower West Pubnico—as opposed to just West Pubnico, or to East Pubnico across the Argyle Sound—is a bucolic little village on Nova Scotia's south shore. I mean, I have lived in bucolic little villages—indeed, I live very near one now—but even by these standards, Lower West Pubnico is very definitely not urban, or industrial. Just a co-op store, a fish plant, a rather good restaurant that serves that curious Acadian dish called rappie pie, and half a hundred houses, in a very good state of repair. The inhabitants fish for a living, and they do well, mostly from lobstering, though they seem constantly to complain about the paucity of the pickings. They are, after all this time, still mostly Acadian in origin, and they share names like Amirault, Belliveau, de'Entremont, and d'Eon; no fewer than three pages of Pubnico's six in the telephone directory are filled with d'Entremonts, with first names ranging from Ada to Yvon. The tallest building in town by far is the church. Or at least it was. Now the church steeple is dwarfed by a series of gigantic windmills, or wind turbines. Their blades alone, rotating at a stately pace in the fresh breeze, are longer than the church is high.

You can see these turbines from across the bay at East Pubnico. They don't in fact look very large, or at all intrusive, but this is because you get no sense of their scale. You can keep them in sight as you round the head of the bay and head back down to West Pubnico. You can see them from the village there, but the curious thing is that they don't look any larger, or smaller, than they did around the bay, although you have traveled a good six miles. They actually look like normal windmills from almost ten miles and they take much longer to get to than you would expect. You travel through the village down the highway to Pubnico Point, which is where the locals used to go for romantic trysts, to watch the odd moose in the swamp, or to test out their ATVs. Finally, the turbines seem to get larger and larger, and if you drive out to where the bulldozers have been grading access roads, you can park your car pretty much underneath the turbine towers (security is not a major concern at Lower West Pubnico). They loom overhead, gigantic and otherworldly, as massive and as unexpected as an office tower in the wilderness. Indeed, these things are the size of an office tower. The hub of the rotor is 257 feet above the ground, and the overall height, blades included, is 389 feet, about the size of a forty-story building. Even the blades are huge, each 262 feet in diameter, 11.62 feet at their widest, with a 13 degree twist. When I first saw them, they were rotating at a leisurely fifteen revolutions a minute, but they are so big that the blade tip was traveling at well in excess of 100 miles an hour. You can get no sense of this at all, until you watch the blade shadow whipping by on the ground, faster than an eyeblink.

I had seen earlier versions of wind turbines, in California and Maine and other places; the wind farm in the Altamont Pass east of San Francisco consists of more than six thousand of the things, of assorted vintages and designs, and I remember that many of them made a variety of more or less unpleasant noises—clanking and creaking, sometimes whining, occasionally as loud as an unmuffered lawnmower. The blades at Pubnico made hardly any sound at all. In a light wind you could hear a faint swishing if you stood directly underneath, but if the wind picked up, the sound of the breeze actually drowned the sound of the blades moving, and they appeared completely silent.

As of spring 2005, fifteen 1.8-megawatt Vestas turbines from

Denmark had been installed. The project's financing was an example of what was becoming a familiar pattern in such green projects. Part of the ownership is local—sensible developers always try to head off on-the-ground opposition by getting the locals involved, and one of the d'Entremonts, Brad, is part owner. The rest is venture capital, money that flows in partly because of an assured customer base; Nova Scotia Power, the provincial generating company, has guaranteed a certain price for a kilowatt hour—a price made possible by a complicated series of incentives and tax breaks, part of the Canadian drive to meet its Kyoto commitments. By 2004 NSP, a notorious coal burner, was getting about 10 percent of its energy from renewables, mostly hydropower, and was looking to increase that to about 25 percent by 2006. About 100 gigawatt hours will come from Pubnico once the wind farm is complete. It would be the largest wind farm in the Canadian Maritime Provinces, with a nominal output of some 30.6 megawatts and an annual production of 100 million kilowatt hours of energy. According to the promotional material cheerfully handed to all and sundry by the builders, this would be "enough to prevent the production of 90,000 tons of CO_2 and 50 tons of NO_2 annually, roughly the equivalent of not driving 16,000 cars or planting 750,000 trees for 60 years. It would be enough energy to supply some 13,000 homes." Nice round numbers, these, but they should be treated with caution.

I spent an hour or so poking about the turbines, being regaled with statistics by an eager maintenance engineer, a local lad who had been taken off to Vestas headquarters in Denmark for training. He reeled off the numbers: the foundation of each tower is 15 feet in diameter and 30 feet deep, the bottom section anchored by 30-foot, two-inch-thick steel bolts spaced every foot around the tower, inside and out. The bottom section of the tower is 37 feet long, with a diameter of 13.2 feet tapering to 12, and weighs 48 tons. The uppermost of four sections that are bolted together is itself 80 feet long and weighs 43 tons. The nacelle where the rotor is housed weighs 68 tons, the rotor another 39 tons. The nacelle is the size of a school bus, and is large enough inside to jump up and down on its floor without hitting the roof. For the brave or exceptionally fool-hardy, at the very top is a sunroof, with an apparently magnificent view—I took his word for it. You get to the top by ladder, 270 feet straight up. And yet so finely balanced is the whole structure that a single technician can haul on a cable and turn the whole massive thing by hand.

I'm dwelling rather longer on Lower West Pubnico and its gen-erating capacity than might seem justified, partly because the mere fact that such a facility, with its sophisticated engineering and com-plex but by-now-familiar financing pattern, has made it to a part of the world that is hardly an industrial powerhouse is a good indica-tion of the wind rush that is consuming the energy industry. Wind farms are being built everywhere from Point Reyes to Nantucket, from the Gulf of Mexico to Wisconsin, from the interior of British Columbia to the craggy coasts of Newfoundland. And in Europe, which is far ahead of America in these matters—pretty well every-where. But wind power is not without its opponents, or its share of controversy. And not without a generous dollop of hype too.

W indmills were among the earliest tech-nologies that replaced humans and domesticated animals as a source of energy, probably after waterwheels, which were easier to fabri-cate than windmills. Where the first windmills were built is, at least judging from the confident but contradictory available sources, still obscure. Some reference books suggest windmills were operating in China by 2000 B.C., but scant evidence has been found for this as-sertion. The U.S. Department of Energy maintains that "by 200 B.C., simple windmills in China were pumping water, while vertical-axis windmills with woven reed sails were grinding grain in Persia and the Middle East." There's no real evidence for these dates either. The first actual historical reference to windmills was from Persia in A.D. 644, but no drawings of the device survive. The first sketches date from 950, and show millers in the Persian city of Seistan grinding grain on a vertical axis windmill. By the eleventh century people all over the Middle East were using windmills extensively for food production. Some reports say the Crusaders brought the idea back to Europe at about that time, but this is doubtful. The very different design of the European mills, generally built on a horizontal axis, implies that they were invented independently. It does seem clear that Persian millwrights, captured by the invading forces of Genghis Khan, were sent to China to construct windmills there, mostly to draw water for irrigation projects on the dry plains north of Beijing.

Once domesticated in Europe, windmills spread rapidly. By the fourteenth century almost all mills everywhere in Europe were taxed, sometimes severely, but this didn't stop their spread. By the eighteenth century there were windmills in nearly every field in Europe. In Britain alone, it is now estimated there may have been almost ninety thousand of them. They simply became part of the countryside. In Holland too. The Dutch were among the preeminent millers, using windmills not just for agriculture and industry but also for draining the lakes and marshes of the Rhine delta. By the early eighteenth century the profile that is still familiar from Dutch landscape paintings appeared all over the continent: a squarish building with a section that could rotate into the wind, with four huge, clanking wooden sails. The Zaan River region of the Netherlands became a global industrial powerhouse, a center of heavy industry and a major exporter, all the factories powered by the wind, the oil of its time.

No one regarded these edifices as eyesores. They were utilitarian devices and lacked any charm except for a marked efficiency, but people grew fond of them for what they represented. As the industrial revolution proceeded apace, and as steam engines gradually replaced windmills for milling and drawing water, nostalgia replaced need, and by the 1990s only one of the Zaan windmills was left, courtesy of local pride and heavy subsidies. A few others survive elsewhere, also as historical curiosities or tourist attractions. One relic is in Cape Town, and I poked through it as a boy, fascinated as all boys seem to be with its array of pulleys and levers and wooden gear wheels. I remember its keeper saying that the wind was only sufficient one day a week, and that the mill even then operated for a couple of hours a day, the rest of the time taken up with trimming its sails and repairing the frequent breakdowns. Another such mill survives in the United States, in the Michigan town of Holland, an authentic Dutch mill built in the 1720s, taken to the United States in 1964.

In the colonial era, windmills soon spread to the most arid of places, in many cases making the difference between viable farming and penury. I remember them from when I was a boy. It hardly rained where I grew up, though when it did, the clouds burst, and for most of the year the rivers were dry, dusty places where thornbushes grew and weaverbirds made their intricate nests. My grandfather only had water courtesy of a borehole he had drilled 950 feet into the shale and rock of the substrata, into an aquifer left over from prehistory, and it was drawn to the surface by a clanking windmill, a mechanical, charmless thing. For a while I thought it *made* the water, somewhere in its rusty heart. Windmills just like it became critical to human spread in the nineteenth century; they made life in the South African barrens possible, they opened up the Australian hinterlands, they followed the dispersal of Americans westward through the dusty High Plains and allowed them to settle there with their stock, in what would otherwise have been a desert. A little later, in the first decades of the twentieth century, windmills not only drew water; they generated what little electricity a household needed. My grandfather had a small Delco wind generator on a tower above his house, which charged up a single battery, enough to fire up the primitive radio (not much better than a crystal set, as I remember), which the family used in the 1940s to gather news of the ominous doings in Europe.

Windmills still clank in remote places of the American West, drawing water without supervision and hardly needing maintenance, far from the grid or any homestead. But in most places the useful windmills have disappeared; replaced at first by the steam engine and then by electricity. In America the Rural Electrification Administration's programs brought inexpensive electric power to most areas in the United States in the 1930s.

The very industrialization that killed traditional windmills also laid the foundations for their further development. Windmills may have been vanishing everywhere, but they had been around recently enough that many scientists remembered them, and the memory set off a generation of inspired tinkerers. As the hunger for electricity increased, so did the notion of capturing the perpetual motion machine of the world's winds, not for direct power as in the past, but to generate power that could then be used for other purposes.

The first wind turbines, as they came to be called, appeared in Denmark in the 1880s. Still, the first windmill built expressly to generate electricity was built by a mechanical engineer, Charles Brush, in Cleveland in 1888. Before him, few had dared to grapple with it, *Scientific American* opined in 1890, "for the question not only involved the motive power itself and the dynamo, but also the means of transmitting the power of the wheel to the dynamo, and apparatus for regulating, storing and utilizing the current."

Brush was, to use a labored pun commonly used to describe him at the time, a dynamo. He was one of the founders of the American electricity industry, for the company he founded merged with Thomas Edison's business under the name General Electric Company.

His wind turbine is now largely forgotten, except to cultural historians. It was a great clanking thing fifty-six feet in diameter, with no fewer than 144 rotor blades made of cedar, and developed a measly 12 kilowatts.[16]

Brush's system used solenoids to control the power output, a technology that didn't change until the 1980s, when computers took over the task. But otherwise his device was soon superseded; the so-called wind-rose design, a large wheel with many blades, was inherently inefficient, and it was a Dane, Poul la Cour, who made the next breakthrough. He built his first models in a wind tunnel and discovered that faster rotation, using many fewer blades, was much better at generating electricity than the slow-moving windmills adapted by Brush. La Cour's first prototype was a sleek, four-bladed machine that turned extremely rapidly.

La Cour was trained as a meteorologist, not an engineer, but he was an inveterate tinkerer and tried out many of his devices in Askov, the community where he and his wife lived. He gave courses on electrical generation at the local high school, published the first-ever wind journal, *The Journal of Wind Electricity*, and founded the Society of Wind Electricians in 1905, three years before his death. In one way he was more farsighted than many who succeeded him, because he recognized one of the drawbacks of wind power, its inherently intermittent nature, and tried to solve the problem by using the electricity he produced to run electrolysis experiments aimed at producing hydrogen to light the gas lamps at the local school. Good idea, except that his engineering skills were not as keen as his imagination, and he had to replace the windows of the school buildings more than once after they blew out—oxygen had leaked into the hydrogen, causing several nice explosions.

By 1918 more than a hundred local utilities in Denmark had at least one wind turbine, usually putting out no more than 20 or 35 kilowatts. In total, wind already accounted for about 3 percent of Danish electricity consumption. Most of these turbines were locally owned and operated, often by cooperatives of farmers, and sited close to the devices they were designed to power. This wide distribution, with the concomitant wide acceptance, was a major reason why Denmark remained the world leader in wind turbines by the year 2005. At the turn of the millennium an astounding 5 percent of the Danish population owned at least one share in a wind turbine, and the Danish public came to see them as natural parts of the landscape, both urban and rural; and their relatively small scale, absent the grandiosities of the first American incarnations, meant that it was easy to make small, incremental, and cumulative design improvements.

By the 1940s, the largest wind turbine ever attempted was built by a Vermonter, Palmer Cosslet Putnam, on a hillside called Grandpa's Knob near his home. It had two seventy-foot stainless steel blades each weighing eight tons, and generated enough power, 1.25 megawatts in a 30 mile an hour wind, to power about two hundred homes, fed through the local utility. In 1945 a blade tore loose, demolishing a few trees along its arc, and the turbine was never repaired. Cheap electricity produced by cheap coal and cheap oil put most of the research efforts on hold.

Until the 1970s, when OPEC's first oil shock hit, setting off another wave of R&D in both America and Europe.

A wind turbine is the opposite of a fan. Instead of using electricity to make wind, wind turbines use wind to make electricity. The wind turns the blades, which spin a shaft, which connects to a generator and makes electricity. So far so simple. What makes it effective is the force that wind generates.

As described earlier, the force exerted by the wind on a structure varies by the square of its wind speed—that is, the force exerted by a 24-mile-per-hour wind is four times the force exerted by a 12-mile-per-hour wind. It's not just wind that does this—the ratio applies to all kinetic motion. If you double the speed of a traveling car, say, it will take four times the power to bring it down to a standstill, in accordance with Newton's second law of motion. But the wind's energy as it applies to windmills is greater still—it increases with the cube (the third power) of its velocity, not the square. That is, if the wind speed is twice as fast, it contains eight times as much energy ($2 \times 2 \times 2$). This apparently puzzling fact is explained by the fact that the wind passes through the turbine, so that if the speed of the wind doubles, twice as many slices of wind pass through the turbine each second, and each slice contains the usual four times as much energy, yielding up the eight-times result.

The speed of the winds, then, is critical, and has great practical implications. Take an average wind speed for a day of an arbitrary 11.2 miles an hour. Say the wind in London blows all day at that speed, 11.2 miles an hour, but the winds in Paris blow at 8.2 miles an hour half the day and 14.2 the other half. The averages are the same, but in practice Paris would get as much power from half a day as London did the whole day. The shorter but faster winds add enormously more power.[17] In electrical generation terms, a wind speed of 26 feet a second will yield a power output of 314 watts for every three feet of turbine; but at 54 feet a second, the power output is 2512 watts.[18]

Utility-scale turbines are almost all 50 kilowatts or larger, up to four megawatts. Single small turbines below 50 kilowatts, often in association with photovoltaic systems, are used for homes, telecommunications dishes, or water pumping. And this is the green utopian dream: for every building a small wind plant and solar panel system, generating a supply of site-specific hydrogen that will run pretty well everything now run electrically. No more grid. No more massive power stations. No more nuclear. No more transmission lines to go down in storms. No more being beholden to OPEC. No more carbon-based pollutants. No more terrorist threats against infrastructure . . . A convert to the dream in 2004 was the mayor of Chicago, Richard Daley, who vowed to turn Chicago into the greenest city in America by 2006. Half of that green power would come from "super modern windmills" whirring silently inside cages encrusted with solar panels, installed atop the city's buildings.[19] This distributed-model dream is also shared by the nuclear industry. Why not a tiny reactor in every building? The two sides don't talk to each other very much.

The oil shock had several other consequences. The most obvious was to push government research grants radically upward. In the late 1960s, U.S. subsidies were a paltry $60,000 a year; six years later they had reached $20 million. Then subsidies were awarded to those who actually produced electricity that could be fed into the grid, subsidies that basically guaranteed producers a profit and allowed the utilities to offset a portion of their costs. The result was a proliferation of wind power companies, and the springing up of wind farms everywhere. The first, and one of the largest, has been part of the landscape on the Altamont Pass, on Interstate 580 east of San Francisco, for thirty years. Another is in Texas; Texans on I-10 heading west to El Paso will see a huge array of turbines stretching endlessly across the plain. In American fashion, entrepreneurship was to be unfettered by overzealous regulation, with consequences that horrified the tidy Danes. The wind farms that sprang up in California, particularly, were, not to put too fine a point on it, ugly. The turbines were of random sizes and random designs; they were often sited on skylines within view of residential areas. In at least one case, the desert community of Palm Springs, the residences they were in sight of were those of the megarich, like Bob Hope, and the squawking was loud and unremitting. Worse, the Palm Springs turbines used a lattice design for their towers, which made them resemble the pylons used to carry power lines across the country. To make things still worse, it seemed that half of them weren't working at any one time. It was a curiosity that all the opinion polls taken at the time were more negative about idled turbines than about working ones.

The Altamont Pass is a dispiriting place for anyone who wants to believe that wind generation of electricity can exist in harmony with nature. I was last there in the spring of 2004; it was a golden California day and I could see apparently forever, wave after wave of silky, golden brown folds in the landscape, dotted here and there with acacia and live oak. In the far distance the air was smoky, a blue haze melting into the golden grass. But in the near distance . . . A passing security patrol let me through a gate; he was bored and pleased to have something, anything, to do, and tramping a complete stranger through the six thousand or so turbines was better than nothing. It was not a pretty sight. A steady wind was blowing, but at least a third of the turbines were idle. Of those that did turn, many were scraping and squeaking and clanking, badly in need of maintenance. Others had toppled altogether. The ground was littered with debris, twisted struts, broken blades, piles of concrete. Access roads had been carelessly scraped into the fragile landscape, and were now eroding into a pattern of ugly scars. There was trash everywhere. Many of the structures housing the controllers and transformers were in need of paint.

If you were really careful, you could angle yourself to screen out the worst of the vistas. If you walk along the hiking trail that traverses the site for about a mile, you can see a ridge with a graceful line of new turbines soaring over it, as elegant as the flight path of a gull, a sight that lifted the heart, but the overall impression was overwhelmingly depressing. "There are some good people with good machines here," said the security guard, squatting on his heels, staring down into a gully. He sounded glum. "But lots of them don't care. I guess they took their money and ran. Too bad they don't make them clear up after themselves." He himself lived down in the valley, within sight of a few hundred turbines. "They don't look too bad," he said, "from a distance."

Altamont is everyone's worst-case scenario. Even on the Sussex Downs in England, wind farm opponents trot out pictures of Altamont to scare the locals.

But the turbines that are being built now almost all follow the Danish model. They are tall, solid towers, no longer made of iron lattice but white painted steel, soaring into the sky, their three-bladed rotors as graceful as storks, as I had seen for myself at Lower West Pubnico. And they are springing up everywhere. The wind rush, indeed, is on.

As an industry, wind energy didn't really exist twenty years ago, but there was a fivefold increase (487 percent) in wind-delivered electricity between 1995 and 2001. Wind power is now growing by 30 percent a year. Windswept Scotland has come to consider itself the future Saudi Arabia of the wind business. Wind farms have been constructed in France, Germany, Holland, Poland, and well into Russia. Offshore wind farms can now be found off virtually every available European coast; Ireland has announced plans to become an offshore power generating megapower. Britain will have four thousand turbines by 2006. In Germany, there were already seven thousand installed by 2004, and the number was increasing rapidly. By Lester Brown's calculations at his WorldWatch Institute, by 2004 enough power was already being extracted from the wind to meet the needs of 23 million people, or the combined populations of the Scandinavian countries. Denmark gets more than a quarter of its energy from wind, a target only dreamed of in North America. Meanwhile coal generation has dropped 9 percent. Wind energy has come of age, to quote the title of the best-known book on wind generation,[20] and has far out-stripped its rivals in the renewable energy department, photovoltaic solar power, tidal power generation, and the like. It is the first re-newable other than hydroelectric power to achieve commercial vi-ability; modern turbines approaching two megawatts are competitive with coal and nuclear power, and better yet, wind power is not vulnerable to a cartel—winds blow everywhere. In the last decade production costs for wind power have dropped from about thirty cents per kilowatt hour to less than six, and some companies have reached three cents, which compares increasingly favorably to the standard two to five cents for conventional fuels. Wind potential is about five times the current global consumption of energy, and can be produced from areas that are not environmentally sensitive.[21]

And some of the heaviest of corporate hitters are rapidly getting into the picture. These include Shell, Scottish Power/PPM Energy, and AES Corp. In the United States, American Electric Power (AEP), the largest United States generator of electricity and a leading coal miner, has gone into wind power in a big way. AEP's Trent Mesa wind farm near Abilene, Texas, has one hundred turbines generating 1.5 megawatts each. Early turbines in Texas were built by Zond Energy Systems, which later became Enron Wind; when Enron went spectacularly belly up, AEP snapped up the assets. Chevron-Texaco power and gasification division head James Houck said in 2004 that "wind power is an increasingly viable source of power generation." Ronald Lehr, a former Colorado public utilities commissioner, said, "The big players who didn't give a hoot four years ago are finally getting into the game, which is precisely what is needed to make wind a viable energy source." Even George Bush, as governor of Texas, signed a bill requiring utilities to get 2,000 megawatts of electricity from renewables by 2009, setting off the largest annual increase in wind power in U.S. history.[22]

A total of ten offshore wind farms were operating in 2004, in Denmark, Sweden, the U.K., and Holland, including the world's largest, Horns Rev in Denmark, at 160 megawatts. The Irish government, not to be outdone, approved plans for an even bigger offshore wind farm, to be built on a sandbank in the Irish sea off Dublin. It would produce 520 megawatts. By early 2005, plans had been announced for twenty-six more, in the U.K., Ireland again, Belgium, Germany, and the Netherlands, and possibly the United States. Britain alone was planning fifteen giant offshore farms, in the Thames estuary, the Wash, and off the northwest and Welsh coasts. Each could have up to five hundred of the biggest turbines available, each turbine generating 4.75 megawatts. Offshore is favored by developers despite the hazards and difficulties of engineering structures to withstand ocean gales, partly because wind shear is low at sea, and less turbulent, and so turbines can be built less tall for the same gain, and are likely to have a longer life. Utility companies had announced investments of some $17.5 billion in British wind projects.[23] Estimates were that about 5 gigawatts of the projected worldwide total of 60 gigawatts by 2010 would come from offshore farms.

Global capacity of wind power was 23,300 megawatts in 2002, and increased by 30 percent each of the following years.

Heady days, then, full of promise but with just a hint, a faint whiff, of economic bubble and hype.

Everywhere, in the United States, or Britain, or anywhere in Europe, wind power is being pushed by government subsidies. Whether this is a good thing depends on which side you have chosen to believe. Opponents come very close to implying the whole thing is a scam; among them the nuclear industry, which has a vested interest in seeing wind power fail—many a nuclear plant is lying around with nothing to do. Wind proponents point to the massive subsidies that oil and gas exploration companies have received over the decades, never mind the nuclear industry, and are aghast at the hypocrisy that now opposes subsidies for a competing technology. That the subsidies do make a difference, no one doubts. Katherine Seelye reported in the *New York Times* that U.S. federal subsidies allow wind power companies to deduct 1.8 cents tax liability for every kilowatt hour they produce for ten years. Jerome Niessen, president of NedPower, which has received West Virginia permission for a two hundred-turbine wind farm in Grant County, said he expected to generate 800 million kilowatt hours a year, for a tax savings of $16 million a year for 10 years, or $160 million on a wind farm that will cost $300 million to build.[24]

Wind power has generated huge controversy in the environmental movement. Sometimes the opposition verges on hysteria, and the picture painted of wind farms is that of some alien monster marching across the countryside, ruining the landscape, killing the wildlife, making life a misery for everyone. It sometimes sounds as though the worst excesses of the industrial revolution are threatening to overwhelm the pristine countryside, as if windmills brought with them belching smokestacks, miles of concrete and asphalt, awful noises, and visual pollution. The perpetrators are portrayed as typical capitalist rapists, representatives of massive multinationals, unconcerned with ordinary people, prepared to blight the world for corporate profit.

The reality, as I have seen for myself, is rather different. Most wind power companies are small start-ups with impeccably green credentials. But sometimes the wind industry has been disingenuous in its claims, slippery with its facts, and with an apparently inbuilt propensity to exaggerate and lie.

The two most interesting examples of how rancorous the debate can get are in Britain and the United States.

The British example, described by John Vidal in *The Guardian,* is from a remote area (well, as remote as you can get in a small island with seventy million people) on the Welsh border, near the magnificent scenery of the Snowdonia mountains. The Conway Valley in Wales is sheep-farming country, and the people who set off the uproar are hardly typical representatives of Big Corporatism. They were sheep farmers themselves, Geraint Davies and the brothers Robin and Rheinalt Williams. They have brought down on their unsuspecting heads a posse of heavyweight greeners, including the Snowdonia Society, the Campaign for the Protection of Rural

Wales, the tourism industry, various rambler societies (a curious British phenomenon dedicated to keeping as much of the country-side open to hikers as possible), the National Trust, the local Labour MP, and various members of the Welsh Assembly. More than sixty national and local groups are signing up for the fight. The wind farm they are objecting to consists of three slender turbines project-ing 150 feet above a derelict stone barn. But yes, they can be seen from some of Snowdonia's peaks.

As *The Guardian* pointed out, few local people objected when the project began. The reasons for it seemed sound. They liked the idea of some of their own diversifying out of sheep farming, which no longer made anyone any money anyway. As Davies put it at the time, "Our copper, our slate, our young people and our water have all gone over the border. Well, our wind won't." He dismisses his critics as white settlers, a nasty dig in Britain, comparing them to the whites who settled in Rhodesia, shamelessly exploiting the black inhabitants. They were rich urbanites who had paid a lot of money for a view and just wanted to protect their investments, he declared. The opposition called wind power lunacy, asserting that it would wreck the environment it was claiming to save, and compar-ing it to the fatuous boast once made by a U.S. military commander in Vietnam that he'd had to destroy a village in order to save it from the enemy.[25] It was NIMBYism taken to an extreme, to protect a yard most of the owners only saw on weekends.

The debate, if it can be dignified by that term, over a proposed 130-turbine wind farm in Nantucket Sound off Cape Cod was uglier still. At one point there was even an indictment, when one of the leaders of the Alliance to Protect Nantucket Sound, the leading antiturbine contender, was charged with planting a fake newspaper article designed to discredit the project's builders as fraud artists. The builder was Jim Gordon, president of Cape Wind Associates, whose plan was to spend $700 million to build America's first off-shore wind farm. His engineers had not picked Nantucket Sound because they wanted to irritate a lot of very wealthy people. They picked it because they needed shallow water, protection from Atlantic storms, isolation from main shipping channels, easy access to the electrical grid and, of course, wind—with an annual average of 18 miles an hour. The sound, in the federal waters of Horseshoe Shoal separating Cape Cod from Martha's Vineyard and Nantucket and less than seven miles from the Kennedy family compound, was the ideal spot.

Among the heavyweight Cape Cod vacationers who opposed the project on aesthetic grounds were a former CEO of a large copper mining company, an attorney who represented Exelon Generation, one of the largest fossil-fuel-generating companies in the United States, Walter Cronkite, and Robert Kennedy Jr. "Our national treasures should be exempt from industrialization," Cronkite put it in a radio broadcast, while rather sheepishly admitting to the *New York Times* that yes, his own house happened to look out at that very national treasure.

The Kennedy connection really grated on those Greens who were for the proposal. The Natural Resources Defense Council, the environmental organization for which Robert Kennedy is a senior attorney, has strongly supported offshore wind power in the past, but here he was, arguing vigorously against just such a project—because, they suspected, he could see it from his front yard. This was not NIMBYism, it was NIVOMDism, "Not in View of My Deck–ism."[26] "I am all for wind power," Kennedy insisted in a debate with the developer on Boston's NPR affiliate. "The costs . . . on the people of this region are so huge . . . the diminishment to property values, the diminishment to marinas, to businesses . . . People go to the Cape because they want to connect themselves with the history and the culture. They want to see the same scenes the Pilgrims saw when they landed at Plymouth Rock." The *New York Times* writer who recounted his rambling defence, Elinor Burkett, pointed out, rather gently, I thought, that the Pilgrims never saw

Nantucket Sound, and if they had, they wouldn't have spied the Kennedy compound. As for the pristine sound being desecrated by a skyline of flashing lights, other project proponents were even more sardonic: "The Sound is not pristine," says Matt Patrick, a member of the state legislature whose support for the plan greatly compromised his reelection campaign. "You can't get to shore because it is lined with memorials to bad taste. Motorboats race around it, and if you go offshore in the summer, you look back and see yellow brown haze hanging over the mainland. And they make it sound as if Nantucket Sound will look like downtown New York, but the wind farm will be only a thumbnail on the horizon." Dick Elrick, a Barnstable councilman who has been a ferryboat captain for two decades, is even angrier, mostly about the support given to the antis by commercial fishermen who themselves operate draggers. "It's tough to listen to the same fishermen who have hurt the habitat by overdragging the bottom of the Sound [now] waving the flag of environmentalism," he said.[27]

The truth of the matter is . . . that the truth of the matter is very hard to discern, a common conclusion in this sort of debate. Both sides seem to be talking at cross purposes. The antis are not antienvironmentalism, for the most part. They are just suspicious of this particular path to salvation. And so they tend to talk in code, and just end up sounding hypocritical. Wind power enthusiasts, on the other hand, are just as disingenuous, and their numbers are not to be trusted. The wattage output numbers in their press releases and announcements are never actuals, or averages, but always maximums attainable only if the machines were working twenty-four hours a day at full capacity in ideal winds. Consequently, the numbers trotted out for saving so many tons of carbon dioxide are always grotesquely inflated; and so are the figures for the numbers of trees that would otherwise have to be planted, and for the homes that each turbine can safely run. It is generally wise to discount the given pronouncements by at least 50 percent, perhaps more. Nor is wind power saving consumers any money. Nor, indeed, are they yet making any money for their operators, except via state subsidies.

On subsidies, though, both sides are right. Wind power is getting government aid, but then it is also true that the oil, gas, and especially the nuclear industry have collectively received subsidies far greater than the total cost of building all the wind plants so far operating in the world.

And on pricing, it is true that wind's prices are competitive with fossil fuels, taking into account those subsidies. With these further advantages: no price spikes by nasty cartels, and no chance of running out of fuel. And the chance either to get off grid and be independent, something that appeals to environmentalists and conservatives alike, or to make a little money selling power to the distribution companies, something to appeal to the small-business person with entrepreneurial instincts. For example, early in 2004, two privately owned turbines were operating in Maine. One, operated by Larry Beaulieu, of Beaulieu in Madawaska, Aroostok County, sold its power to Maine Public Power. It was a tiny 0.05 megawatts. The other was the same size, owned by G. M. Allen and Sons of Deer Island, and powered their own blueberry farm.[28]

In all the places where wind power is opposed, the arguments are similar. They are ugly. They despoil pristine places and beautiful landscapes. They are being built in the wrong places ("here"). They destroy property values and drive away tourists. They are land hungry. They are noisy and dangerous. They put wildlife at risk. They are too expensive anyway and we should be looking for other technologies. They are intermittent and can't be relied on, and therefore ensure that fossil-fuel or nuclear plants would have to be retained as the main generators.

It can be difficult, at times, to pick a valid argument out of the white noise that contaminates the debate, with its wild claims that whales will crash into offshore turbines, that fishing grounds will be destroyed, that dead birds will litter the beaches, that bats die in great numbers near them, even that horses bolt when they hear them, a curiously nineteenth-century argument. Bird kill is a major issue—the turbines are called pole-mounted Cuisinarts by the antis (this claim was started by a California group, which called one of the Altamont turbines a "Condor Cuisinart"). Some of these reports are doubtless true, if exaggerated. It's also true that more birds are killed by guyed towers, which are much less visible to avian eyes but much more prevalent in the landscape, and more than six times as many are killed by domestic cats every week as are killed in a year by all the wind farms put together. (For the record, the actual kill rate for a turbine is 0.2 birds per turbine per year.) Ordinary high-rises kill more birds than that.

The only argument against wind power that has any real merit is that it is, by definition, intermittent and can't be relied on, and therefore we have to keep a substantial investment in fossil-fuel or nuclear-generating plants as backups when the wind fails. Wind power's proponents answer that wind power is not intended to be a stand-alone technology. Allan Moore, chairman of the British Wind Energy Association and head of renewables at a company called National Wind Power, agrees that a mix of technologies will be necessary. Still, he told *The Guardian*, wind is far more advanced than the others. "If in 30 years time someone comes up with something better, we'll take the turbines away." This is not very difficult. A decommissioned turbine can be taken away, and will leave behind only a very small sign of its former presence. Wind power, then, is part of a basket of solutions that include solar power, biomass fuels, tidal and current harnessing, as well as conservation. (The idea of harnessing the awesome power of the moving water of the ocean's currents, especially the mighty Gulf Stream, has been seriously proposed. It has some advantages over tidal power, since it is constant and not intermittent; the Gulf Stream is only a few miles off the Florida shore, where it is moving at a rapid 2.4 miles an hour. Surveys have shown that 400 to 850 gigawatts of energy are plausible from this source, enough to cover the needs of several states. Indirectly, this is wind power too, since ocean currents are wind-driven.)[29]

The argument about wind's intermittent nature also ignores the interesting possibility of hydrogen technology. As Vijay Vaitheeswaran puts it, "In the long term, the world will get its hydrogen directly from renewable energy, whether from the wind or the sun, by electrolysis of water. Once produced, hydrogen would also be used as a form of energy storage. Power generated whenever the wind blows can be stored as hydrogen and sold into the power grid when needed, which would revolutionize the way electricity trading is done, since electricity is one of the few commodities that cannot be stored easily, but must be used as soon as produced."[30]

Moreover, the argument that you can't work around the intermittent nature of wind, or that if you could it would be pointless to do so, rather leaves out the case of Denmark, which already gets 27 percent of its energy from renewables, almost all wind, against a target-to-date of 20 percent. On windy days, the capacity often goes up to 50 percent in the western part of the country. Curiously, this causes problems of its own, quite the reverse of the "how are we going to burn our toast if the wind doesn't blow" conundrum. The problem, rather, is an oversupply.

It arises because much of the rest of Denmark's generating capacity comes from coal-fired thermal generators. These are relatively inflexible—slow to fire up, slow and expensive to spin down. Under normal loads, there is always need for a reserve of backup power; this reserve is usually about 20 percent, or in the Danish case,

about the size of the single largest coal-fired plant. It is used to balance the variations in the wind output, and to counterbalance the
minute by minute variations in demand. But what to do with over-
capacity, or oversupply? What the Danes do is offload the surplus to
neighboring Norway, whose generating system is 99 percent hydro,
which is fairly easy and relatively cheap to spin down. Norway takes
the surplus, relieving the pressure on the Danish system.

It all comes down to costs. But how to measure the true costs?
By annual revenues accruing to operators? By tax revenues forgone?
Or by some more indirect measurement? What are the true costs of
building turbines in pristine places? Costs to the serenity of the environment, to scenic beauty, to quality of life? To the value of privately owned land that abuts wind farms? What are the costs of not
building them? How to measure the costs to the environment of
not heading off climate change and global warming if we can?

"You can't have your cake and eat it too." Gary Gallon, who
writes a newsletter for the Canadian Institute for Business and the
Environment, cheerfully trots out the cliché. "You can't say no coal,
no oil-fired electricity, unless you can provide an alternative, more
benign energy source."

As Elinor Burkett put it: "To [environmentalists] the national illusion that you can have electricity, clean air, a stable climate and independence from foreign oil without paying a steep price is
ludicrous. In fact, in late April [2003], part of the price Cape Cod is
already paying began washing up on its shores. En route to a power
plant in Sandwich, on the northwest corner of the Cape, a leaking
barge spilled 98,000 gallons of oil into Buzzards Bay. Shellfish beds
were closed for a month. At least 370 birds died; 93 miles of coastline were tainted by thick globs of black oil."[31]

None of the windpower proponents could be caught saying "I
told you so." They are far too media savvy for that. But an undercur-
rent of complacency nevertheless crept into their communications
afterward. They clearly felt that the dirty calculus of industrialization

would continue to do their work for them, providing for the contemplation of the electorate endless evil examples of pollution and degradation of the environment. After a while—you could hear them thinking—we won't have to sell wind power anymore. Its need will become self-evident.

Epilogue

*I*van's dangerous death throes and its perversely *complicated demise: By the time Ivan crossed the New Jersey shore into the Atlantic, it had been reduced to a posttropical disturbance, and the U.S. Hurricane Center lost interest, discontinuing their public advisories. They were turning their attention to Hurricane Karl, then apparently heading harmlessly (except of course to mariners) into the central Atlantic, and tropical storms Jeanne and Lisa, either of which could develop a temperament as nasty and unpredictable as Ivan's.*

But, as it turned out, consigning Ivan to the archives was a little premature. In the movies, what happened next would either be called Ivan 2, *or,* The Return of Ivan. *Because the monster had a few surprises in store yet.*

Ivan had seemed to drift out into the ocean, but that was only half the story. Quite literally, because Ivan split in two. It was the lower half that drifted out to sea. It was still spinning slowly, but was now below the radar of the forecasters, in both senses of the phrase. The satellites and Doppler radars ignored what was happening, and the forecasters had more serious things on their minds. Lisa had behaved even more oddly than usual for a tropical storm, and was now heading eastward instead of westward, away from the Caribbean and North America. The satellites kept a still-wary eye on her, but paid more attention to Jeanne, already a hurricane and heading, alas, for Florida. The midlevel high that had prevented the month's storms from their normal northerly recurvature had broken up, and Jeanne could easily pound Florida and then head up the coast. The configuration of the jet stream was such that it was possible—not likely, but possible—that Jeanne could race up the coast and intersect with Maritime Canada. That was enough to catch northern attention. Even had anyone been paying attention to the leftover Ivan, they would not have taken it seriously.

This part of Ivan, call him Low Ivan, drifted slowly southward in a leisurely clockwise circle, and ended up—where else this year?—on the Florida coast, as it turned, once again, westward. The winds were moderate by tropical cyclone standards—not much more than 20 to 25 miles an hour—although there was a fair amount of precipitation. This low drifted across the Florida peninsula into the Gulf of Mexico. There, like an old warhorse smelling action, it encountered the warm water of the Gulf, was reenergized, and took on the familiar organization characteristics of a tropical depression—spinning a little faster, warm moisture-laden air ascending, high-altitude cold convection currents, accelerating winds . . .

The Hurricane Center wryly admitted on the evening of September 22 that there had been "considerable and sometimes animated in-house discussion of the demise [or supposed demise] of Ivan. In the midst of a low-pressure and surface frontal system over the eastern United States . . . the National Hurricane Center has decided to call the tropical cyclone now over the Gulf of Mexico Tropical Depression Ivan. While debate will surely continue here and elsewhere . . . this decision was based primarily on the reasonable continuity observed in the analysis of the surface and low-level circulation." Whatever the name, satellite images and buoys in the Gulf showed that the disturbance was organized enough to be called a tropical depression, and the low level of the shear indicated to the forecasters that the depression might very well become a tropical storm by landfall, expected sometime along the Texas Gulf coast. A tropical storm warning was issued for the Gulf coast from the mouth of the Mississippi in Louisiana to Sargent, Texas.

This was a nasty surprise to the Texans, who had been relieved to see

Ivan pass by them to their east a week earlier, and had no wish to reprise what Alabama and Florida had then suffered.

As it turned out, the Hurricane Center forecast was a little pessimistic. Ivan did cross the Gulf coast near Cameron, Louisiana, but with winds that seldom exceeded 30 miles an hour, even in gusts, and was weakening rapidly. It turned toward Texas, passed over the town of Port Arthur before turning southwest, and finally sighed to a halt near the coast on the early morning of Sunday, September 26. There was no need to drive a spike through its heart. It simply expired.

So much for half the story, Low Ivan. What of High Ivan?

Peter Bowyer takes up the story: "The upper half of Ivan was picked up in the prevailing southwesterlies and flew up into eastern Canada. It was not strong enough to produce damaging winds on its own, but coincidentally a frontal system coming in from the west was developing. That storm would have happened anyway—we would have had winds gusting to perhaps 80, 90 kilometers, but again, on its own not enough to cause any real damage. But the two of them together . . . Leftover Ivan become entrained in this frontal system, and the result was like taking a fire and throwing kerosene into it."[1] Ivan's rotation and massive quantities of Caribbean moisture were the kerosene; the frontal system the fire. The combination was enough to turn the new system—Ivan Redux—into a weather bomb, which as we have seen is a slightly hysterical though still technically rigorous term, defined as a system that is already at less than 1,000 millibars when it drops 24 further millibars in twenty-four hours. On September 21, this storm exceeded those criteria by more than half, and turned the system back into the equivalent of a Category 1 hurricane that roared across the northern Nova Scotian mainland, the islands of Cape Breton and Newfoundland, uprooting trees, flooding roads, leaving more than 18,000 people without power for several days, and killing six mariners. The massive cruise ship Queen Mary, *which had been scheduled to dock in the northern port of Sydney that day, departed hastily for open ocean and more southerly latitudes; the ferry to Newfoundland was knocked out of service; roads were covered with debris, and all schools were closed.*

*O*n *the afternoon of September 21 the sea in front of our house was crashing on the rocks. The swells had been building all day, and by four o'clock immense rollers were breaking on the bedrock with a great roar. A small spruce tree outside my office window was smacking against the eaves with an unnerving scraping sound. A larger spruce was swaying alarmingly against the power lines coming in from the highway almost a mile away, and we prepared once more for a power outage. The weathervane was pointing northeast; we measured gusts at 50 miles, then 60, then 65. The house, strongly built though it is, creaked and the shutters banged. A skylight moaned in the wind, a ghost-wind squeezing through some minuscule hole.*

I didn't know that was Ivan, then. For me, Ivan was a killer whose narrative was a series of printed bulletins, now safely contained in a file folder. He was just a story. *He wasn't supposed to hit my house.*

It could have been worse, of course; in very many places, it was.

*W*hen we bought our house down on the shore, here at the end of the peninsula, it had attached to it a venerable wind turbine on a forty-foot tower. The fellow who built the house was what they call a "belt and suspenders" kind of guy; he heated his house with wood but had a gasoline generator and electric heaters as backup; used the wind turbine to feed a bank of car batteries to keep his lights going, but kept several kerosene lanterns just in case. Later, he even hooked up to the grid for good measure. At some point before we bought the place, a lightning strike had fried the batteries; the windmill had been disconnected from the house and clanked disconsolately in the wind. Even when there was no wind at ground level, every now and then it would abruptly start up and rattle around for a minute or two before spinning down. We had it removed soon after we bought the property. The buyer came with a crane and trucked it away, and we installed the fish we called Wanda as a weathervane on the windmill's tower.

Now we're thinking of buying a new wind generator. They are smaller than they were a few years ago, lighter, more reliable, easier to use, and are responsive to gentle winds as well as to greater ones. Our house is too far from Lower West Pubnico to benefit from that wind farm's power generation, and in any case, Pubnico is pumping its power directly into the grid, and we don't really trust the stability and security of the grid anymore.

This lack of trust is expressed through a basket of concerns about increasingly erratic weather, the security of long-term fuel supplies, and the uncertainties that greenhouse gases represent for the global climate. These are in effect worries not just about our own place in the scheme of things but also for the scheme of things itself—that is, for the planet's future; and both sets of concerns were neatly encapsulated for me by the monster, Ivan, which was not only a uniquely intense storm, but traveled more than eight thousand miles and caused enormous damage and loss of life, and then, before expiring, cast its baleful eye on me personally, and took a swipe at my house.

Ivan didn't batter me into the sea as the gale in Cape Town had threatened to do, or even batter the sea into my house, though it was plenty strong enough. Perhaps as a consequence of the possibilities, I remain wary of wind but would like to get a little of my own back and put this wariness to some use, to harness wind to increase my own comfort and security. In this, I think, I am also Us, in the larger sense of a collectivity and a species. We are badly wounding our air and through it our climate, but we know enough now to be able to bring the system back to health. We know how wind works, and what makes it worse and what doesn't, and how to make it cleaner and less hazardous to our survival. We have also learned how to employ it to generate the energy that our civilization needs. Nature has given us the perpetual motion machine we call wind. We can put it to work to make things better.

APPENDIX 1

The composition of the modern atmosphere

Component	Symbol	Content (% of total)
Nitrogen	N_2	78.084
Oxygen	O_2	20.947
Argon	Ar	0.934
Carbon dioxide	CO_2	0.035 (370 parts per million)
Neon	Ne	18.2 parts per million
Helium	He	5.2 parts per million
Krypton	Kr	1.1 parts per million
Sulfur dioxide	SO_2	1.0 parts per million
Methane	CH_4	2.0 parts per million
Hydrogen	H_2	0.5 parts per million
Nitrous oxide	N_2O	0.5 parts per million
Xenon	Xe	0.09 parts per million
Ozone	O_3	0.07 parts per million
Nitrogen dioxide	NO_2	0.02 parts per million
Iodine	I_2	0.01 parts per million
Carbon monoxide	CO	trace
Ammonia	NH_3	trace

The atmosphere contains gases that are considered to be permanent (which remain essentially constant by percent) and gases considered to be variable (which have changing concentrations over a finite period of time).

Permanent gases:
Nitrogen	78.1%
Oxygen	20.9%

Argon	0.9%
Neon	0.002%
Helium	0.0005%
Krypton	0.0001%
Hydrogen	0.00005%

Variable gases:

Water vapor	0 to 4%
Carbon dioxide	0.035%
Methane	0.0002%
Ozone	0.000004%

APPENDIX 2

The Beaufort scale
The original scale, as devised by Beaufort himself

1	Light air	Or just sufficient to give steerage way
2	Light breeze	Or that in which a man-of-war with all sail set, and clean full would go in smooth water from 1 to 2 knots
3	Gentle breeze	Or that in which a man-of-war with all sail set, and clean full would go in smooth water 3 to 4 knots
4	Moderate breeze	Or that in which a man-of-war with all sail set, and clean full would go in smooth water 5 to 6 knots
5	Fresh breeze	Or that to which a well-conditioned man-of-war could just carry in chase, full and by Royals, &c.
6	Strong breeze	Single-reefed topsails and top-gal. sail
7	Moderate gale	Double-reefed topsails, jib, &c.
8	Fresh gale	Treble-reefed topsails &c.
9	Strong gale	Close-reefed topsails and courses
10	Whole gale	Or that with which she could scarcely bear close-reefed main-topsail and reefed fore-sail
11	Storm	Or that which would reduce her to storm staysails
12	Hurricane	Or that which no canvas could withstand

The first nonmariner's version of the Beaufort scale

0	Calm	Calm; smoke rises vertically
1	Light air	Direction shown by smoke but not by wind vanes
2	Light breeze	Wind felt on face; leaves rustle; vane moved by wind
3	Gentle breeze	Leaves and small twigs in constant motion; wind extends light flag
4	Moderate breeze	Raises dust and loose paper; small branches are moved
5	Fresh breeze	Small trees in leaf begin to sway; crested wavelets on inland waters
6	Strong breeze	Large branches in motion; telegraph wires whistle; umbrellas used with difficulty
7	Near gale	Whole trees in motion; inconvenience in walking against wind
8	Fresh gale	Breaks swigs off trees; generally impedes progress
9	Strong gale	Slight structural damage occurs; chimney pots and slates removed
10	Whole gale	Trees uprooted; considerable structural damage occurs
11	Violent storm	Very rarely experienced; accompanied by widespread damage
12	Hurricane	Devastation occurs

The modern Beaufort scale

Beaufort number	Wind speed (in knots)	Description
0	<1	Calm
1	1–3	Light air

Beaufort number	Wind speed (in knots)	Description
2	4–6	Light breeze
3	7–10	Gentle breeze
4	11–16	Moderate breeze
5	17–21	Fresh breeze
6	22–27	Strong breeze
7	28–33	Near gale
8	34–40	Gale
9	41–47	Strong gale
10	48–55	Storm
11	56–63	Violent storm
12	>64	Hurricane

APPENDIX 3

The Saffir-Simpson hurricane scale

Category 1:
Sustained winds 74–95 mph (64–82 knots). Storm surge generally 4–5 feet above normal. No real damage to building structures. Damage primarily to unanchored mobile homes, shrubbery, and trees. Some damage to poorly constructed signs. Also, some coastal road flooding and minor pier damage.

Category 2:
Sustained winds of 96–110 mph (83–95 knots). Storm surge generally 6–8 feet above normal. Some roofing material, door, and window damage. Considerable damage to shrubbery and trees with some trees blown down. Considerable damage to mobile homes, poorly constructed signs, and piers. Coastal and low-lying escape routes flood 2–4 hours before arrival of the hurricane center. Small craft in unprotected anchorages break moorings.

Category 3:
Sustained winds 111–130 mph (96–113 knots). Storm surge generally 9–12 feet above normal. Some structural damage to small residences and utility buildings with a minor amount of curtainwall failures. Damage to shrubbery and trees with foliage blown off trees and large trees blown down. Mobile homes and poorly constructed signs are destroyed. Low-lying escape routes cut by rising water 3–5 hours before arrival of the center of the hurricane. Flooding near the coast destroys smaller structures with larger structures damaged by battering from floating debris. Terrain lower than 5 feet above sea level may be flooded inland 8 miles or more.

Evacuation of low-lying residences within several blocks of the shoreline may be required.

Category 4:
Sustained winds of 131–155 mph (114–135 knots). Storm surge generally 13–18 feet above normal. More extensive curtainwall failures with some complete roof structure failures on small residences. Shrubs, trees, and all signs blown down. Complete destruction of mobile homes. Extensive damage to doors and windows. Low-lying escape routes may be cut by rising water 3–5 hours before arrival of the center of the hurricane. Major damage to lower floors of structures near the shore. Terrain lower than 10 feet above sea level may be flooded requiring massive evacuation of residential areas as far inland as 6 miles.

Category 5:
Sustained winds greater than 155 mph (135 knots). Storm surge generally greater than 18 feet above normal. Complete roof failure on many residences and industrial buildings. Some complete building failures with small utility buildings blown over or away. All shrubs, trees, and signs blown down. Complete destruction of mobile homes. Severe and extensive window and door damage. Low-lying escape routes are cut by rising water 3–5 hours before arrival of the center of the hurricane. Major damage to lower floors of all structures located less than 15 feet above sea level and within 500 yards of the shoreline. Massive evacuation of residential areas on low ground within 5–10 miles of the shoreline may be required.

APPENDIX 4

Hurricane strikes in the U.S.

U.S. hurricane strikes by decade

Decade	Saffir–Simpson Category					Total	Major
	I	2	3	4	5		
1900–1909	5	5	4	2	0	16	6
1910–1919	8	3	5	3	0	19	8
1920–1929	6	4	3	2	0	15	5
1930–1939	4	5	6	1	1	17	8
1940–1949	7	8	7	1	0	23	8
1950–1959	8	1	7	2	0	18	9
1960–1969	4	5	3	2	1	15	6
1970–1979	6	2	4	0	0	12	4
1980–1989	9	1	5	1	0	16	6
1990–1999	3	6	4	0	1	14	5
1900–1999	61	39	48	14	3	165	65

Note: "Major" is Category 3 through 5.
Source: U.S. National Hurricane Center

U.S. hurricane strikes by state, 1900 through 1996

	I	2	3	4	5	Total	Major
Texas	12	9	9	6	0	36	15
(North)	7	3	3	4	0	17	7
(Central)	2	2	1	1	0	6	2
(South)	3	4	5	1	0	13	6
Louisiana	8	5	8	3	1	25	12
Mississippi	1	1	5	0	1	8	6
Alabama	4	1	5	0	0	10	5
Florida	17	16	17	6	1	57	24

	1	2	3	4	5	Total	Major
(Northwest)	9	8	7	0	0	24	7
(Northeast)	2	7	0	0	0	9	0
(Southwest)	6	3	6	2	1	18	9
(Southeast)	5	10	7	4	0	26	11
Georgia	1	4	0	0	0	5	0
South Carolina	6	4	2	2	0	14	4
North Carolina	10	4	10	1	0	25	11
Virginia	2	1	1	0	0	4	1
Maryland	0	1	0	0	0	1	0
Delaware	0	0	0	0	0	0	0
New Jersey	1	0	0	0	0	1	0
New York	3	1	5	0	0	9	5
Connecticut	2	3	3	0	0	8	3
Rhode Island	0	2	3	0	0	5	3
Massachusetts	2	2	2	0	0	6	2
New Hampshire	1	1	0	0	0	2	0
Maine	5	0	0	0	0	5	0
U.S.	58	36	47	15	2	158	64

Source: U.S. National Hurricane Center

Major hurricane direct hits on the mainland U.S. coastline from 1900 to 1996, by state and month

AREA	June	July	Aug.	Sept.	Oct.	Total
Texas	1	1	7	6	0	15
(North)	1	1	3	2	0	7
(Central)	0	0	1	1	0	2
(South)	0	0	3	3	0	6

AREA	June	July	Aug.	Sept.	Oct.	Total
Louisiana	2	0	4	5	I	12
Mississippi	0	I	I	4	0	6
Alabama	0	I	0	4	0	5
Florida	0	I	2	15	6	24
(Northwest)	0	I	0	5	I	7
(Northeast)	0	0	0	0	0	0
(Southwest)	0	0	I	5	3	9
(Southeast)	0	0	2	7	2	II
Georgia	0	0	0	0	0	0
South Carolina	0	0	0	3	I	4
North Carolina	0	0	2	8	I	II
Virginia	0	0	0	I	0	I
Maryland	0	0	0	0	0	0
Delaware	0	0	0	0	0	0
New Jersey	0	0	0	0	0	0
New York	0	0	I	4	0	5
Connecticut	0	0	I	2	0	3
Rhode Island	0	0	I	2	0	3
Massachusetts	0	0	0	2	0	2
New Hampshire	0	0	0	0	0	0
Maine	0	0	0	0	0	0
Total	2	3	15	36	8	64

Source: U.S. National Hurricane Center

APPENDIX 5

Canadian tropical cyclone statistics

Only two major hurricanes (Saffir-Simpson Category 3 or above) have ever made landfall in Canada: an unnamed Category 3 storm in 1893 that made landfall in St. Margaret's Bay, Nova Scotia; and Hurricane Luis, also a Category 3 storm, which made landfall on the Avalon Peninsula, Newfoundland, in 1995.

Hurricane Juan, which hit Nova Scotia and Prince Edward Isand in 2003, was first classified as a Category 1 storm, but it was on the margins of a Category 2, and has been so reclassified.

The worst storms, ranked by intensity (winds in knots)

Year	Name	Pressure*	Wind	Rating	Month	Location
1995	Luis	963	105	SS3	9	NF
1893	1893C	–	100	SS3	8	NS
1963	Ginny	–	90	SS2	10	PEI, NS, NB
1927	1927A	–	90	SS2	8	NS
1908	1908B	–	85	SS2	8	NS
1891	1891F	–	85	SS2	10	NS, NF
2003	Juan	973	80/85	SS1/2	9	NS, PEI
2002	Gustav	–	70	SS1	9	NS
2000	Michael	–	75	SS1	10	NF
1893	1893E	–	88	SS1	8	NF
1924	1924B	–	80	SS1	8	PEI, NS, NF
1896	1896E	–	80	SS1	10	NS

Year	Name	Pressure*	Wind	Rating	Month	Location
1891	1891I	–	80	SS1	10	PEI, NS, NF
1891	1891D	–	75	SS1	9	NS, NF
1969	Gerda	–	70	SS1	9	PQ, NF
1937	1937G	–	70	SS1	9	PEI, NF, NS
1936	1936O	–	70	SS1	9	NS
1893	1893D	–	70	SS1	8	PQ, NF
1892	1892B	–	70	SS1	8	NF
1996	Hortense	980	65	SS1	9	NS
1971	Beth	–	65	SS1	8	NS, NF
1966	Celia	–	65	SS1	7	PQ, NF, NS, NB
1962	Daisy	–	65	SS1	10	NS
1958	Helene	968	65	SS1	9	NF
1940	1940E	–	65	SS1	9	NF, NS
1935	1935A	–	65	SS1	8	NF
1926	1926B	–	65	SS1	8	NS, NF
1924	1925C	–	65	SS1	9	NS
1887	1887E	–	65	SS1	8	NF
1990	Bertha	978	60	TS	8	NF, NS
1975	Blanche	988	60	TS	7	NS
1962	Ella	–	60	TS	10	NF
1959	unnamed	–	60	TS	6	NF, NS, PEI
1954	Hazel	–	60	TS	10	ON
1953	Carol	–	60	TS	9	PQ, NB
1949	1949D	–	60	TS	9	NF
1939	1939D	–	60	TS	8	PQ, NF
1893	1893I	–	60	TS	10	ON, PQ, NF

Year	Name	Pressure*	Wind	Rating	Month	Location
1998	Earl	964	55	TS	9	NF
1979	David	986	55	TS	9	NB, NF
1964	Dora	–	55	TS	9	NF
1933	1933M	–	55	TS	9	NS
1933	1933T	–	55	TS	10	NS
1908	1908D	–	55	TS	9	NF
1904	1904B	–	55	TS	9	NS, NF
1996	Bertha	995	50	TS	7	NS, NF, PEI, NB
1985	Ana	996	50	TS	7	NF
1985	Gloria	987	50	TS	9	PQ
1979	Subtrop1	982	50	TS	10	NF
1973	Alice	–	50	TS	7	NF
1968	Gladys	–	50	TS	10	NS
1957	Audrey	–	50	TS	6	ON
1954	Carol	992	50	TS	9	PQ, NB
1946	1946D	–	50	TS	9	NF
1935	1935D	–	50	TS	10	NF
1926	1926D	–	50	TS	9	NF
1924	1924C	–	50	TS	9	NF
1923	1923B	–	50	TS	10	NF
1893	1893F	–	50	TS	8	PQ, NB, NF
1888	1888I	–	50	TS	10	NS, NF

Note: TS = Tropical Storm; SS1 = Saffir-Simpson Category 1 Hurricane; SS2 = Saffir-Simpson Category 2 Hurricane, etc.

★ A dash (–) under Pressure means that it was not measured.

Key: NB = New Brunswick NF = Newfoundland NS = Nova Scotia ON = Ontario PEI = Prince Edward Island PQ = Quebec

Most active years of landfalling tropical cyclones in Canada

Year	Number
1893	6
1996	5
1995	4
1937	4
1923	4
1891	4
1888	4
1988	3
1979	3
1954	3

Most common date for a landfalling hurricane in Canada

September 15th

Average number of tropical cyclones affecting Canada each year

From 1901 to 2000: 3.3
From 1951 to 2000: 4.2
From 1993 to 2002: 4.9
Source: Canadian Hurricane Centre

APPENDIX 6

World's worst tropical cyclones (hurricanes and typhoons, by year, with casualties)

Location	Year	Casualties
United States	1900	6,000 (Galveston storm)
China	1912	50,000
China	1922	60,000
United States	1938	600
Asia	1942	61,000
Japan	1959	5,098 (Typhoon Vera)
Bangladesh	1959	14,000
Bangladesh	1960	10,000
Bangladesh	1963	22,000
Bangladesh	1965	19,280
United States	1965	75 (Hurricane Betsy)
Bangladesh	1970	300,000
Honduras	1974	9,000 (Hurricane Fifi)
India	1977	14,000
United States	1983	21 (Hurricane Alicia)
Bangladesh	1985	11,050
U.S. & Mexico	1988	355 (Hurricane Gilbert)
U.S. & Mexico	1989	86 (Hurricane Hugo)
Bangladesh	1991	139,000
Japan	1991	62 (Typhoon Mireille)
United States	1992	62 (Hurricane Andrew)
India	1998	10,000
U.S. & Mexico	1998	4,000 (Hurricane Georges)
U.S. & Mexico	1998	9,200 (Hurricane Mitch)

Location	*Year*	*Casualties*
U.S. & Mexico	1999	61 (Hurricane Floyd)
Asia	1999	29 (Typhoon Bart)
India	1999	15,000

Source: Munich Reinsurance Company

APPENDIX 7

Wind speed variation within the hurricane eyewall, by elevation

Height (in feet)	Height (in stories)	Wind (% surface)	Pressure force (% surface)
33	3	100	100
50	5	103	106
100	10	108	117
150	15	111	123
200	20	115	132
250	25	117	137
300	30	119	142
400	40	121	146
500	50	123	151
600	60	125	156
750	75	128	164
1,000	100	131	172

Source: U.S. National Hurricane Center

APPENDIX 8

Worst winter storms on record

Location	Year	Casualties
Europe	1953	1,932 (storm surge)
Germany	1967	40
Europe	1972	54
Europe	1976	82 (Winter Storm Capella)
United States	1982	270
Europe	1982	48
United States	1983	500
Europe	1987	17
Europe	1990	230
United States	1992	19
U.S. & Canada	1998	45 (ice storm)
U.S. & Canada	1999	25
Spain	1999	
Europe	1999	20 (Winter Storm Anatol)
Europe	1999	100 (Winter Storm Lothar)
Europe	1999	30 (Winter Storm Martin)

Source: Munich Reinsurance Company

APPENDIX 9

The Fujita tornado scale

Fujita 0, gale tornado:
Winds 40—72 mph. Such tornadoes cause some damage to chimneys, break branches off trees, and push over shallow-rooted trees.

Fujita 1, moderate tornado:
Winds ranging from 74 to 112 mph. The lower limit of a moderate tornado is the sustained wind speed that defines a Category 1 hurricane. Such tornadoes can peel off roofs, overturn mobile homes, and push cars off roads. Some poorly made buildings will be destroyed.

Fujita 2, significant tornado:
Winds ranging from 113 to 157 mph. Such winds will do considerable damage, tearing roofs off many houses, demolishing mobile homes, snapping large trees. "Light-object missiles" will be generated—debris picked up in the winds that become battering rams.

Fujita 3, severe tornado:
Winds from 158 to 206 mph. Roofs and walls torn off well-made buildings, trees uprooted, and even trains overturned.

Fujita 4, devastating tornado:
Winds ranging from 207 to 260 mph. In these conditions even well-made houses are leveled. Structures with weak foundations will be blown some distance. Cars are thrown about, and "large missiles" generated.

Fujita 5, incredible tornado:

Winds of 261 to 318 mph, Strong frame houses lifted off their moorings, car-sized missiles flying about, trees debarked, steel-reinforced concrete badly damaged.

Fujita 6, inconceivable tornado:

Sustained winds of 319 to 379 mph, but no one will ever know, because all measuring devices would be destroyed, along with pretty well everything else. (The Fujita scale recognizes that "the small area of damage they might produce would probably not be recognizable along with the mess produced by F4 and F5 winds that would surround the F6 winds. Missiles, such as cars and refrigerators would do serious secondary damage that could not be directly identified as F6 damage. If this level is ever achieved, evidence for it might only be found in some manner of ground swirl pattern, for it may never be identifiable through engineering studies.")

Source: tornadoproject.com and the National Severe Storms Laboratory

APPENDIX 10

Worst tornadoes in the twentieth century by year

Location	Year	Casualties
United States	1925	739
United States	1936	455
United States	1965	271
Germany	1968	3
Bangladesh	1969	500
United States	1974	320
Bangladesh	1977	900
United States	1979	48
United States	1982	46
United States	1984	80
United States	1985	94
Canada	1987	26
South Africa	1990	2
United States	1991	33
Bangladesh	1996	600
India	1998	200
United States	1998	1
United States	1999	51

Source: Munich Reinsurance Company

APPENDIX 11

Wind force table

For a house whose face presents an area of 400 square feet, we can predict the following approximate lateral inertial forces as the wind speed increases:

Wind speed (mph)	Inertial force (lb.)
20	400
40	1,600
60	3,600
80	6,400
100	10,000
120	14,400
140	19,600
160	25,600

Source: Ernest Zebrowski Jr., *Perils of a Restless Planet*

APPENDIX 12

Canadian wind chill index (in degrees Celsius and wind speeds in kilometers per hour)

Air Temperature

Wind speed values (km/h) are listed in the left column. The left margin spells **WIND SPEED** vertically.

Wind Speed	16	14	12	10	8	6	4	2	0	-2	-4	-6	-8	-10	-12	-14	-16	-18	-20	-22	-24	-26	-28	-30	-32	-34	-36	-38	-40
4	16	14	12	10	8	6	3	1	-1	-3	-6	-8	-10	-12	-14	-17	-19	-21	-23	-26	-28	-30	-32	-35	-37	-39	-41	-43	-46
6	16	14	12	10	9	7	3	0	-2	-4	-7	-9	-11	-14	-16	-18	-20	-23	-25	-27	-30	-32	-34	-37	-39	-41	-43	-46	-48
8	16	14	11	9	7	4	2	0	-3	-5	-7	-10	-12	-14	-17	-19	-22	-24	-26	-29	-31	-33	-36	-38	-40	-43	-45	-47	-50
10	16	13	11	9	6	4	1	-1	-3	-6	-8	-10	-13	-15	-18	-20	-22	-25	-27	-29	-31	-34	-36	-39	-41	-44	-46	-49	-51
12	16	13	11	8	6	3	1	-1	-4	-6	-9	-11	-13	-16	-18	-21	-23	-26	-28	-30	-33	-35	-37	-40	-42	-45	-47	-50	-52
14	15	13	10	8	6	3	1	-2	-4	-7	-9	-12	-14	-16	-19	-21	-24	-26	-29	-31	-34	-36	-38	-41	-43	-46	-48	-51	-53
16	15	13	10	8	5	3	0	-2	-5	-7	-10	-12	-15	-17	-20	-22	-24	-27	-29	-32	-34	-36	-39	-41	-44	-47	-49	-52	-54
18	15	13	10	8	5	3	0	-2	-5	-7	-10	-12	-15	-17	-20	-22	-25	-27	-30	-32	-35	-37	-40	-42	-45	-47	-50	-52	-55
20	15	12	10	7	5	2	0	-3	-5	-8	-10	-13	-15	-18	-20	-23	-25	-28	-30	-33	-36	-38	-41	-43	-46	-48	-51	-53	-56
22	15	12	10	7	5	2	0	-3	-6	-8	-11	-13	-16	-18	-21	-23	-26	-28	-31	-34	-36	-39	-41	-44	-46	-49	-51	-54	-56
24	15	12	10	7	4	2	-1	-3	-6	-8	-11	-13	-16	-19	-21	-24	-26	-29	-31	-34	-37	-39	-42	-44	-47	-49	-52	-54	-57
26	15	12	9	7	4	2	-1	-3	-6	-9	-11	-14	-16	-19	-22	-24	-27	-29	-32	-34	-37	-40	-42	-45	-47	-50	-52	-55	-58
28	14	12	9	7	4	2	-1	-4	-6	-9	-11	-14	-17	-19	-22	-25	-27	-30	-32	-35	-37	-40	-43	-45	-48	-50	-53	-56	-58
30	14	12	9	7	4	1	-1	-4	-6	-9	-12	-14	-17	-20	-22	-25	-27	-30	-33	-35	-38	-40	-43	-46	-48	-51	-53	-56	-59

(continued)

Appendix 12 (*Continued*)

Air Temperature

Wind Speed	16	14	12	10	8	6	4	2	0	-2	-4	-6	-8	-10	-12	-14	-16	-18	-20	-22	-24	-26	-28	-30	-32	-34	-36	-38	-40
32	14	12	9	6	4	1	-1	-4	-7	-9	-12	-15	-17	-20	-22	-25	-28	-30	-33	-36	-38	-41	-43	-46	-49	-51	-54	-57	-59
34	14	12	9	6	4	1	-2	-4	-7	-10	-12	-15	-17	-20	-23	-25	-28	-30	-33	-36	-39	-41	-44	-46	-49	-52	-54	-57	-60
36	14	11	9	6	4	1	-2	-4	-7	-10	-13	-15	-18	-20	-23	-26	-28	-31	-34	-36	-39	-42	-44	-47	-49	-52	-55	-57	-60
38	14	11	9	6	3	1	-2	-5	-7	-10	-13	-15	-18	-21	-23	-26	-29	-31	-34	-37	-39	-42	-44	-47	-50	-52	-55	-58	-60
40	14	11	9	6	3	1	-2	-5	-7	-10	-13	-15	-18	-21	-23	-26	-29	-31	-34	-37	-39	-42	-45	-48	-50	-53	-56	-58	-61
42	14	11	9	6	3	0	-2	-5	-8	-10	-13	-16	-18	-21	-24	-26	-29	-32	-34	-37	-40	-42	-45	-48	-51	-53	-56	-59	-61
44	14	11	8	6	3	0	-2	-5	-8	-10	-13	-16	-18	-21	-24	-27	-29	-32	-35	-37	-40	-43	-45	-48	-51	-54	-56	-59	-62
46	14	11	8	6	3	0	-2	-5	-8	-11	-13	-16	-19	-21	-24	-27	-30	-32	-35	-38	-40	-43	-46	-48	-51	-54	-57	-59	-62
48	14	11	8	6	3	0	-3	-5	-8	-11	-13	-16	-19	-22	-24	-27	-30	-32	-35	-38	-41	-43	-46	-49	-51	-54	-57	-60	-62
50	14	11	8	5	3	0	-3	-5	-8	-11	-14	-16	-19	-22	-24	-27	-30	-33	-35	-38	-41	-44	-46	-49	-52	-54	-57	-60	-63
52	14	11	8	5	3	0	-3	-6	-8	-11	-14	-16	-19	-22	-25	-27	-30	-33	-36	-38	-41	-44	-47	-49	-52	-55	-57	-60	-63
54	14	11	8	5	3	0	-3	-6	-8	-11	-14	-17	-19	-22	-25	-28	-30	-33	-36	-39	-41	-44	-47	-50	-52	-55	-58	-60	-63
56	13	11	8	5	2	0	-3	-6	-9	-11	-14	-17	-20	-22	-25	-28	-31	-33	-36	-39	-42	-44	-47	-50	-53	-55	-58	-61	-64
58	13	11	8	5	2	0	-3	-6	-9	-11	-14	-17	-20	-23	-25	-28	-31	-34	-36	-39	-42	-45	-47	-50	-53	-56	-58	-61	-64
60	13	11	8	5	2	0	-3	-6	-9	-12	-14	-17	-20	-23	-25	-28	-31	-34	-36	-39	-42	-45	-48	-50	-53	-56	-59	-61	-64

From -25° to -34°: Frostbite likely after prolonged skin exposure to wind

From -35° to -60°: Frostbite possible in less than 10 minutes

Below -60°: Frostbite possible in less than 2 minutes

ACKNOWLEDGMENTS

And many thanks to Peter Barss, Peter Bowyer, Dereck Day, Chris Fogarty, Tamara Gates-Hollingsworth, Bill Gilkerson, Nicole Meoli, Don Sedgwick and Shaun Bradley, Bruce Whiffen, and the forecasters of the national hurricane centers in Miami, Florida, and Dartmouth, Nova Scotia.

NOTES

Note: For full publishing data of the books cited, see bibliography.

CHAPTER ONE

Wind's Mystery and Meaning

Information in this chapter comes from a variety of sources; most are cited in the text. Particularly useful were David E. Newton's quirky but informative *Encyclopedia of Air*, Sebastian Smith's lyrical memoir of sailing in the Mediterranean, *Southern Winds*, and Jan DeBlieu's *Wind*, which covers much of the same ground as this volume. DeBlieu brings a poet's eye to her meditations on wind.

[1] Newton, *Encyclopedia of Air*, pp. 132–33, as quoted in *Golden Bough*. From *Encyclopedia of Air* by David E. Newton, copyright © 2003 by David E. Newton, reproduced with permission by Greenwood Publishing Group, Westport, CT.

[2] Smith, *Southern Winds*, p. 100. From *Southern Winds*, by Sebastian Smith, copyright Sebastian Smith 2004, Penguin, London.

[3] Dorson, *Buying the Wind*, pp. 32–33.

[4] Quoted in Newton, *Encyclopedia of Air*, p. 133.

[5] From the Flat Earth Society Web site, an elaborate spoof compiled by "Lee Harvey Oswald Smith." www.cloudwall.com.

[6] *Larousse Encyclopedia of Mythology*, p. 144.

[7] Smith, *Southern Winds*, p. 30.

[8] These are the "named winds" listed on Whirling Winds of the World (www.cloudwall.com): Abroholos, Afternoon Burner, Antane, Aspre, Auster, Austru, Autun, Badisad obistroz, Barat, Barber, Barines, Barrier winds, Belat, Bellot, Berg, Bhoot, Bise, Black South Easter, Blue Norther, Bochorno, Bohorok, Bolon, Bora, Boreas, Bornan, Brickfielder, Bricklayer, Brisote, Broeboe, Bruscha, Buran, Burga, Candlemas Crack, Candlemas Eve, Canterbury Northwester, Cape Doctor, Cers, Chabascos, Chamsin, Chergui, Chinook, Chocolatta, Chocolatero, Coho, Collada, Contrastes, Cook Strait Southerly, Cordonazo, Coromell, Cowshee, Craudelaire, Criador, Crivetz, Descuernaccabras, Diablo, Dog's Tongue, Dusenwind, Dzhani, El Cierzo, Elephanta, Etesian, Etobicoke Echo, Euraquilo, Euroclydon, Eurus, Foeh, Foehn, Fremantle Doctor, Gallego, Galerna, Garbin, Garigliano, Gending, Ghibli, Gharbi, Glaves, Golfada, Greco, Gregale, Grigale, Guba, Gully Squall, Guxen, Guzzle, Haboob, Haize-Beltza, Halny Wiatr, Harmattan, Havgull, Helm, Hora, Hot Busters, Ibe, Jauch, Jauk, Jochwinde, Joran, Junkwinde, Junta, Juran, Kachchan, Kaikias, Kal

Baisahki, Karaburan, Karajol, Kapalilua, Kaus, Kesisleme, Khamsin, Kharif, Kloof, Knik, Kona, Koshava, Kuban, Laawan, Labbe, Lansan, Laventera, Leste, Levante, Levanter, Leveccio, Leveche, Liberator, Lips, Livas, Ljuka, Llebetjado, Lodos, Lombarde, Loo, Luganot, Maestro, Maestrale, Maloja, Marajos, Marin, Matacabras, Matanuska, Mauka, Meltemi, Minuano, Mistral, Mono, Morget, Muscat, Nashi, Nevada, Newall Winds, Night Winds, Nirta, Nortes, Notus, Oberwind, Ora, Orsure, Palouser, Pampero, Panas Oetara, Papagayos, Paramito, Passat, Piner, Piterak, Polacke, Poniente, Ponente, Poriaz, Pruga, Puelche, Puna, Purga, Pyrn, Rachas, Rebat, Reshabar, Robin Hood's Wind, Roger, Rondada, Rotenturm Wind, Rotetur, Safid Rud, Sahel, Samiel, Sansar, Santa Ana, Sarca, Scirocco, Seca, Seistan, Shaluk, Shamal, Sharav, Sharki, Siffanto, Simoom, Simoon, Simoun, Skiron, Sky Sweeper, Sno, Solano, Sonora, South Easter, Southerly Buster, Stikine, Suete, Suhaili, Sukhovey, Sumatra, Sundowner, Sures, Suroet, Take, Takn, Taku, Tamboen, Tauem, Techuantepecer, Temporale, Texas Norther, Tramontana, Turnagain, Vent du Midi, Vendavale, Williwaw, Yamaoroshi, Yildiz, Zephyr, Zephyros, Zoboa, Zonda.

[9]Rigveda, canto 186.

[10]Guy de Maupassant, as quoted in *Southern Winds*, pp. 17, 87.

[11]Conrad, *Typhoon*, pp. 84, 96.

[12]As quoted in *Southern Winds*, p. 107.

[13]Joyce Boro on the BBC's weather channel, part of their Weather in Literature background server.

[14]Smith, *Southern Winds*, p. 194.

[15]Ibid., p. 85.

[16]*A Narrative of Travels in North Africa in the Years 1818, 19, and 20*, by Captain G. F. Lyon, London 1821, p. 94.

[17]*Southern Winds*, p. 113.

[18]Quoted in *Southern Winds*, p. 86.

[19]Joseph d'Agnese, "Why Has Our Weather Gone Wild," *Discover*, June 2000.

[20]DeBlieu, *Wind*, p. 180. Excerpts from *Wind: How the Flow of Air Has Shaped Life, Myth and the Land*, by Jan DeBlieu, copyright 1998 by Jan DeBlieu, reprinted by permission of Houghton Mifflin Company. All rights reserved.

[21]Ibid., p. 181.

[22]D'Agnese, "Why Has Our Weather Gone Wild" (see note 19, above).

[23]Quoted in DeBlieu, *Wind*, p. 181.

[24]Felicity James, on the BBC's weather channel, from her article, "Weather in Literature: The Modern Novel."

CHAPTER TWO

Wind's Great Theater

Richard Fortey's *The Earth: An Intimate History* is a paleogeographic look at the earth from the inside out, and wonderfully well written. Erik Larson's *Isaac's Storm* is a retelling of the hurricane that all but destroyed Galveston, Texas, at the turn of the twentieth century; he is particularly good on the politics of the weather service at the time.

[1] Fens Tian et al, "A Hydrogen-rich Early Earth Atmosphere," *Science*, May 13, 2005, p. 1014.

[2] *Encyclopaedia Britannica*, 15th edition, 14:1155.

[3] Fortey, *The Earth*, pp. 359–61, 366.

[4] Newton, *Encyclopedia of Air*, p. 17.

[5] Larson, *Isaac's Storm*, p. 38.

[6] This history of early chemistry is from a variety of sources. Useful and informative was the chemistry Web site operated by the University of Pennsylvania. www.upenn.edu.

[7] Michael Klesius, "Altitude and the Death Zone," *National Geographic*, May 2003.

[8] Dr. James L. Green, "Magnetosphere," www.ssdoo.gsfc.nasa.gov/education.

CHAPTER THREE

The Search for Understanding

Scott Huler's *Defining the Wind* is a book calculated to appeal to all of us in the writing trade; Huler was a copy editor who became obsessed with Beaufort's mastery of imagery, and it led him to a larger investigation of wind measurement technology. *Hurricane Watch*, by the former director of the National Hurricane Center in Miami, Dr. Bob Sheets, and his collaborator, Jack Williams, is the best source for information on tropical cyclones. Jim Carrier's *The Ship and the Storm* is the story of the fatal encounter between Hurricane Mitch and the charter tall ship *Fantome*.

[1] Quoted in Larson, *Isaac's Storm*, p. 38.

[2] DeBlieu, *Wind*, and many others.

[3] Quoted in DeBlieu, *Wind*, p. 32.

[4] Smith, *Southern Winds*, p. 169.

[5] Pliny, *Natural History*, book 2, c xlvi.

[6] Smith, *Southern Winds*, p. 29.

[7] Ibid., p. 173.

[8]Huler, *Defining the Wind*, p. 82. From *Defining the Wind*, by Scott Huler, copyright 2004 by Scott Huler, used by permission of Crown Publishers, a division of Random House, Inc.

[9]DeBlieu, *Wind*, p. 32.

[10]Anecdotes from Huler, *Defining the Wind,* pp. 62, 64–65.

[11]Larson, *Isaac's Storm*, p. 44.

[12]Huler, *Defining the Wind*, pp. 81, 85.

[13]Quoted in Sheets, *Hurricane Watch*, p. 12. From *Hurricane Watch* by Bob Sheets and Jack Williams, copyright 2001 by Jack Williams and Bob Sheets, used by permission of Vintage Books, a division of Random House, Inc.

[14]Huler, *Defining the Wind*, p. 65.

[15]Huler, *Defining the Wind*, p. 84.

[16]Sheets, *Hurricane Watch*, pp. 13, 14.

[17]DeBlieu, *Wind*, p. 61.

[18]Carrier, *The Ship and the Storm*, p. 148.

[19]*Typhoon,* p. 77.

[20]Sheets, *Hurricane Watch*, Larson, *Isaac's Storm*, and many others.

CHAPTER FOUR

Wind's Intricate Patterns

This chapter owes a debt to the Environment Canada meteorological service, and the Canadian Hurricane Centre, especially the ever-patient Peter Bowyer and Chris Fogarty. Tamara Gates of Environment Canada put me onto Bruce Whiffen's discussions of the Wreckhouse and other local winds.

[1]One knot is 1 nautical mile an hour. A nautical mile is 6076.12 feet or 1852 meters. Therefore 1 knot is 1.152 miles an hour, or 1.85 kilometers an hour.

[2]Smith, *Southern Winds*, pp. 220–21.

[3]Hydrogen atoms weigh 1.08 atomic mass units (AMU); helium atoms weigh 4.0026, yielding a 0.0294 shortfall.

[4]Huler, *Defining the Wind*, p. 60.

[5]Expressed in mathematical language, 1.74×10^{17} watts.

[6]J. W. Chamberlain and D. M. Hunten, *Hadley Circulation, Theory of Planetary Atmospheres* (San Diego: Acadian Press, 1987), pp. 79–80.

[7]In a paper titled "On the Equations of Relative Motion of a System of Bodies."

[8]David J. Van Domelen, of the Ohio State University Department of Physics, gave me my first understanding of the Coriolis force. For the more technically minded, he also gives the following equation relating the Coriolis force to an object's mass (m), its velocity in a rotating frame (v_r), and the angular velocity of the rotating frame of reference (ϖ): $F_{Coriolis} = -2\,m\,(\varpi \times v_r)$. See his Web site at http://www.physics.ohio-state.edu/~dvandom/Edu/newcor.html.

[9]Newton, *Encyclopedia of Air*, p. 131.

[10]Chris Fogarty, "Extratropical Transition of Hurricane Michael," *Journal of the American Meteorological Society,* September 2004, p. 1323.

[11]Chris Fogarty, Canadian Hurricane Centre, Dartmouth, interview.

[12]DeBlieu, *Wind*, p. 5.

[13]*Romance of the Sea*, J. H. Parry, Washington, National Geographic Society, 1981, p. 254.

[14]David Jennings, spokesman for the Canadian Coast Guard, quoted in "*Maersk* Damage," *Halifax Chronicle Herald*, January 28, 2003.

[15]DeBlieu, *Wind*, pp. 64–65.

[16]Peter Bowyer, interview, November 2004.

[17]*Encyclopaedia Britannica* 6:543.

[18]*Encyclopaedia Britannica* 16:454.

[19]Smith, *Southern Winds*, pp. 52–54.

[20]R. E. M. Rickaby and P. Halloran, "Cool La Niña During the Warm of the Pliocene," *Science*, March 25, 2005, p. 1948.

[21]A good exposition of these events is from Josh Larson, USATODAY.com/weather/.

[22]Jury, McQueen, and Levy, *Journal of Theoretical Applied Climatology*, no. 50, pp. 103–15, and Allan, Lindesay, and Parker, *Journal of the American Meteorological Society*, no. 69(2), pp. 24–27.

[23]Sheets and Williams, *Hurricane Watch*, p. 270.

[24]Interview, October 2004.

[25]Kelly Shiers, "Timetable for Next Hurricane," *Halifax Chronicle Herald*, September 25, 2004.

[26]Newton, *Encyclopedia of Air*, p. 135.

[27]Sheets and Williams, *Hurricane Watch*, p. 6.

[28]Vortex material from T. R. Joe Sundaram, who owns and operates an engineering research firm in Columbia, Maryland, at http://www.worldandi.com.

[29]The energy within a single tornado has been calculated at between 10^4 and 10^7 kwh, not much less than the 20-kiloton bomb dropped on Hiroshima, which was 10^{13}.

[30]P. J. Hufstutte, "Tornado Rips Apart Tavern," *Los Angeles Times*, April 22, 2004.

[31]Ibid.

[32]This discussion of tornadoes owes considerable debts to the *Encyclopaedia Britannica* article, "Climate and Weather," vol. 16, p. 436 ff.; to a *National Geographic* magazine article on tornadoes, April 2004; and to Robert Henson and Erik Rasmussen, who authored a piece in *Natural Science*, May 1995 on tornadoes. Henson is the author of *Television Weathercasting: A History*; Rasmussen is a scientist at the National Severe Storms Laboratory in Norman, Oklahoma, and the coordinator of a field program called the Verification of the Origins of Rotation in Tornadoes Experiment (VORTEX).

[33]Trachard, *Voyage to Siam*, p. 34.

[34]*Encyclopaedia Britannica,* "Climate and Weather" (see note 32).

[35]In an essay in *What If,* a compilation of historical might-have-beens edited by Robert Cowley.

[36]Smith, *Southern Winds*, p. 55.

[37]Ibid., p. 135.

[38]DeBlieu, *Wind*, p. 117.

[39]Proulx, *The Shipping News*, p. 317.

[40]Wreckhouse stories and the Lauchie McDougall story by Mont Lingard in *Next Stop: Wreckhouse*, quoted by Bruce Whiffen, November 30, 1998. Used with his permission.

[41]Bowyer, *Where the Wind Blows*, p. 20, and from Tamara Gates, Environment Canada.

[42]Information about Cermak and Davenport from Leighton S. Cochran, associate at Cermak Peterka Petersen, Inc.

[43]Estanislao Oziewicz, "Ivan Brushes Cuba and Heads for the Gulf," *Globe and Mail*, September 14, 2004.

[44]Ibid.

[45]Joy Woller, University of Nebraska–Lincoln Physical Chemistry Lab.

[46]Boundary Layer wind tunnel laboratory study, 2002.

CHAPTER FIVE

The Art of Prediction

Where the Wind Blows is one of those rare publications that delivers more than it promises. It is an admirably clear exposition of weather and its causes edited by Peter Bowyer of the Canadian Hurricane Centre.

[1]The *Fantome*'s demise is the subject of a gripping book by Jim Carrier, *The Ship and the Storm.*

[2]Defoe, *The Storm*, pp. x, 60, 66–67.

[3]Sheets and Williams, *Hurricane Watch*, p. 4.

[4]Bowyer, *Where the Wind Blows*, pp. 30–32.

[5]Smith, *Southern Winds*, p. 11.

[6]Bowyer, *Where the Wind Blows*, p. 6.

[7]Trachard, *Voyage to Siam*, p. 34.

[8]David Kasanof, "The Lore of the Sea," *Wooden Boat* magazine, January-February 2003, p. 13.

[9]Newton, *Encyclopedia of Air*, p. 72.

[10]Robert T. Ryan, *Bulletin of the American Meteorological Society*, 1982, quoted on NASA's earth observatory Web site.

[11]Bowyer, *Where the Wind Blows*, p. 10.

[12]This brief history of weather forecasting is by Steve Graham, Claire Parkinson, and Mous Chahine, from NASA, *The Earth Observatory.*

[13]Beaufort's story is very well told in Scott Huler's *Defining the Wind.* Some of the information about Beaufort here is from that source.

[14]British Met Office, via BBC weather service.

[15]Scott, *Defining the Wind*, p. 48.

[16]Zebrowski, *Perils*, p. 249.

[17]David E. Fisher, *The Scariest Place on Earth: Eye to Eye with Hurricanes*, quoted in Carrier, *The Ship and the Storm*, p. 137.

[18]Scale reproduced from http://www.tornadoproject.com (St. Johnsburg, Vermont).

[19]Newton, *Encyclopedia of Air*, pp. 11–12.

[20]DeBlieu, *Winds,* p. 174.

[21]The formula is: $W = 13.12 + 0.6215\ T_{air} - 11.37\ V_{0.16} + 0.3965\ T_{air}\ V_{0.16}$, where W = the wind chill index, based on the Celsius temperature scale; but note that it is expressed without units, i.e., not with °C. T_{air} = the air temperature in degrees Celsius. V = the wind speed at 10 meters in km/h.

[22]BBC weather service.

[23]BBC weather service.

[24]Sheets and Williams, *Hurricane Watch*, p. 61 ff.; Larson, *Isaac's Storm*, p. 9.

[25]NOAA Web site, October 10, 2003. www.noaa.gov.

[26]From a discussion on the WaterObserver.org Web site by Robert Beal of Johns Hopkins and William Pichel of NOAA.

[27]Sharon Kedar and Frank Webb, Jet Propulsion Laboratory, in *Science*, February 4, 2005, p. 682.

[28]WaterObserver.org.

[29]Ibid.

[30]Bill Power, "Hurricane Sounds Fascinate Acoustics Scientist," *Halifax Chronicle Herald*, December 2, 2003.

[31]Carrier, *The Ship and the Storm*, p. 83.

[32]Interview, November 2004.

[33]Summary of the models by Mark DeMaria, National Hurricane Center, November 26, 1997.

[34]Fogarty interview, 2004.

[35]Carrier, *The Ship and the Storm*, p. 71.

[36]Much of the Herb Hilgenberg material was first published in *Canadian Geographic* magazine, "The Beacon of Burlington," September–October 1997.

CHAPTER SIX

The Most Furious Gale

Apart from *Hurricane Watch*, already cited, Ernest Zebrowski Jr.'s *Perils of a Restless Planet* contains excellent discussions on wind force (cited in the text), hurricane origins, and computer tracking of storms. Wind damage is only one of the perils Zebrowski deals with in his fascinating book, which I highly recommend.

[1]Sheets and Williams, *Hurricane Watch*, p. 16.

[2]Gerry Forbes, head of Sable Island Station, interview with Sheila Hirtle, October 2002.

[3]Paul Doherty, "Blowing in the Wind," December 20, 2001. On the Exploratorium Web site, http://www.exploratorium.edu/origins.

[4] March–April 2000 issue of *Archeology* magazine, quoted by Gareth Cook, "Visions in the Sound: Could it Be that the Pyramids Were Inspired, Not by Aliens, but by the Desert Itself?" *Boston Globe*, May 15, 2001, p. C1.

[5]Herodotus, *Histories*, book 3, p. 26.

[6]Bowyer and MacAfee's paper was called "The Theory of Trapped-Fetch Waves with Tropical Cyclones—An Operational Perspective." The Queen Elizabeth wave was reported by R. W. Warwick, et al, "Hurricane Luis, the Queen Elizabeth II and a Rogue Wave," *Marine Observer*, 1966, 66:134.

[7]Bowyer, *Where the Wind Blows*, p. 48.

[8]William Gilkerson, in *Pirate's Passage*, Shambhala, New York, 2005.

[9]Analogy from Peter Bowyer.

[10]*Encyclopaedia Britannica*, 11:408.

[11]Larson, *Isaac's Storm*, p. 196.

[12]Zebrowski, *Perils*, p. 249.

[13]Ibid., pp. 249–50; and *Encyclopaedia Britannica* 6:738.

[14]*Fine Homebuilding* magazine December 1992–January 1993, p. 82.

[15]Sheets and Williams, *Hurricane Watch*, pp. 215–16.

[16]Applied Research Associates study, quoted by Horia Hangan, July 2002, Institute for Catastrophic Loss Reduction, research paper no. 19. See http://www.iclr.org.

[17]Patent # was 6,601, 348. This information is from Leighton Cochran, from a paper on wind engineering as related to tropical cyclones, "Wind Effects on Lowrise Buildings," from the Web site of the wind consultancy CPP (cppwind.com).

[18]Pacific data from Sheets and Williams, *Hurricane Watch*, pp. 196–97.

[19]Stewart, quoted in Sheets and Williams, *Hurricane Watch*, p. 130.

[20]Carrier, *The Ship and the Storm*, p. 69.

[21]Analysis of computer modeling can be found in Zebrowski, *Perils*, pp. 263 ff.

[22]Sheets and Williams, *Hurricane Watch*, p. 211.

[23]Zebrowski, *Perils*, p. 259.

[24]Sheets and Williams, *Hurricane Watch*, pp. 267.

[25]Interview, November 2004, Las Vegas, Nevada.

[26]Sheets and Williams, *Hurricane Watch*, pp. 98, 99.

[27]Interview, 2004.

[28]Sheets and Williams, *Hurricane Watch*, p. 112.

[29]Carrier, *The Ship and the Storm*, p. 96.

[30]Sheets and Williams, *Hurricane Watch*, p. 107.

[31]Ibid., p. 162.

[32]"Anti-hurricane Technology," *The Economist*, June 11, 2005, Technical Supplement, p. 8.

[33]Zebrowski, *Perils*, p. 274.

[34]Hoffman, material from Sora Song, *Time*, January 1, 2005.

CHAPTER SEVEN

An Ill Wind

The oddly titled *Power to the People*, by *The Economist*'s Vijay Vaitheeswaran, is not at all a lefty political polemic but a fascinating analysis by a liberal economist of the energy industry and its future. There are several citations from the book in this chapter and the next.

[1]Ed Ayres, *Worldwatch* editorial, July–August 2004, p. 5.

[2]Vaitheeswaran, *Power to the People*, p. 164.

[3]C. Venkatamaran, et al., Indian Institute of Technology, reported in *Science*, 4 March 2005, p. 1454.

[4]Ruprecht Jaenicke, "Abundance of Cellular Material and Proteins in the Atmosphere," *Science*, April 1, 2005, p. 73.

[5]U.N. study called "North America's Environment," quoted by Sandra Cordon, "Curb Energy Demands or Expect More Severe Weather UN Warns," *Halifax Chronicle Herald*, August 15, 2002.

[6]Kenneth E. Wilkening, Leonard A. Barrie, and Marilyn Engle, "Trans-Pacific Air Pollution," *Science*, no. 290, October 6, 2000, p. 65.

[7]"Rust Never Sleeps," *The Economist*, December 11, 2004.

[8]Study mentioned on CBC radio *The Current*, September 22, 2004.

[9]John C. Ryan, "Dust in the Wind," *Worldwatch*, January–February 2002.

[10]"From Beijing to Hawaii," *The Straits Times*, Singapore, June 1, 2004.

[11]Gerhard Wotawa and Michael Trainer, "The Influence of Canadian Forest Fires on Pollutant Concentrations in the United States," *Science*, April 14, 2000, p. 324.

[12]Hajime Akimoto, "Global Air Quality and Pollution," *Science*, December 5, 2003, p. 1716.

[13]The studies were: TOMS (Total Ozone Mapping Spectrometer) and SAGE (Stratospheric Aerosol and Gas Experiment), whose instruments on the *Nimbus 7* satellite measured ozone concentrations; GOME (Global Ozone Monitoring Experiment) and SCHIAMACHY (Scanning Imaging Absorption Spectro-Meter for Atmospheric ChartographY), which was studying sodium dioxide, carbon dioxide, and HCHO emissions; MOPITT (Measurement of Pollution in the Troposphere), which studied the worldwide spread of carbon monoxide

pollution; MODIS (Moderate-Resolution Imaging Spectrometer) on the *Terra* satellite, which showed the global distribution of man-made and natural aerosols; and AVHRR (Advanced Very High Resolution Radiometers), a device for remote sensing, which showed the distribution of four major aerosol types, soil dust, and carbonaceous emissions, sulfates, and sea-salt aerosols in the east China Sea. Finally, NASA contributed an aircraft- and shuttle-based study called the Global Tropospheric Experiment, with four separate field missions: Pacific Exploratory Missions A and B, which studied the impact of emissions from Asia over the western Pacific and near the equator in the Atlantic; TRACE-A, which investigated trace gases over the tropical Atlantic; and TRACE-P, which did the same over the Pacific.

[14]Both these quotes, as it happens, were from a single *National Geographic* article on carbon dioxide, "The Case of the Missing Carbon" by Tim Appenzeller, in the February 2004 issue, but there are dozens like them.

[15]Newton, *Encyclopedia of Air*, p. 74.

[16]Vaitheeswaran, *Power to the People*, p. 127.

[17]"Hockey stick" graph by Michael Mann, et al., in *Nature*, no. 392, April 23, 1998, pp. 779–787. See also *Scientific American*, March 2005, p. 34. Hansen's study is available through *Science Express*, April 28, 2005 (http://www.sciencexpress.org).

[18]Pallé et al., "Changes in Earth's Reflectance Over the Past Two Decades," *Science*, May 28, 2004; the second study that disputed its conclusions, in the same magazine, May 6, 2005.

[19]*Bulletin of the American Meteorological Society*, March 2001.

[20]Ed Ayres, *Worldwatch*, November–December 2004, p. 14.

[21]Vaitheeswaran, *Power to the People*, p. 124.

[22]Tim Appenzeller, "The Case of the Missing Carbon," *National Geographic*, February 2004.

[23]Study featured in *Science*, February 25, 2005, p. 1190.

[24]Webster, et al, *Science*, September 2005, p. 1844.

[25]*Hurricane Watch,* p. 267.

[26]Jennifer Kahn, *Harper's*, May 2004, p. 83.

[27]Tim Appenzeller, "The Case of the Missing Carbon," *National Geographic*, February 2004.

[28]David Ebner, "Kyoto Solution?" *Globe and Mail*, September 7, 2004, p. B1.

[29]Environmental Protection Agency report, February 28, 2005.

[30]Kahn, *Harper's* (see ref 26).

[31]Vaitheeswaran, *Power to the People*, p. 167.

[32]Environmental News Service, February 1, 2004.

[33]David W. Keith and Alexander E. Farrell, "Rethinking Hydrogen Cars," *Science*, no. 301, July 18, 2003, p. 315.

[34]Vaitheeswaran, *Power to the People*, p. 113.

[35]The eleven companies were Alcoa; Anglo American, the South African mining house; Cemex, Holsim, and Lafarge, all three cement companies; Hewlett-Packard; the Russian joint stock company Unified Energy System; RWE, a German utility; Scottish Power; Vattenfall, a Swedish energy company; and Vitro.

[36]"The Future Is Clean," *The Economist*, September 4, 2004, p. 61.

[37]Environment News Service, April 12, 2005.

[38]Vaitheeswaran, *Power to the People*, pp. 158–59; and Center for Atmospheric Science Web site, U.K. www.atm.ch.cam.ac.uk/cas/.

CHAPTER EIGHT

The Technology of Wind

The best sources for information on the rapidly changing world of wind power generation is Paul Gipes's *Wind Power Comes of Age*, an admirably dispassionate book from a wind power proponent, and the Web site wind power.org. Vijay Vaitheeswaran's book, cited in the previous chapter, is also good on the potential future of renewable energy.

[1]P. A. Glick, USDA Technical Bulletin #673, 1939. pp. 9, 10

[2]DeBlieu, *Wind*, p. 101.

[3]Ibid., p. 87. See other discussions in the same book on insects and the winds.

[4]Paul Evans, "Aerial Acrobatics," *The Guardian*, February 11, 2005, p. 22

[5]Newton, *Encyclopedia of Air*, p. 88

[6]Ulrike Müller and David Lentink, "Turning on a Dime," *Science*, December 10, 2004, p. 1899, commenting on study by J. J. Videler, et al., published in the same issue.

[7]This and following anecdotes on early aeronautics from Newton, *Encyclopedia of Air*, pp. 5, 125, 153, 179.

[8]Lilienthal data and quotes from a Web-based biography by Gary Bradshaw. www.wam.umd.edu/~stwright/WrBr/taleplane.html.

[9]Michael Klesius, "The Future of Flying," *National Geographic*, December 2003.

[10]Menzies, *1421*, p. 85.

[11]Johnson, *Phantom Islands*, p. 48

[12]Sailor's shanty and clipper ship data from Dyson, *Spirit of Sail*, pp. 11, 17, 18,

21, 23–24, 108, 111.

[13]Heatter, *Eighty Days to Hong Kong*, foreword.

[14]Dyson, *Spirit of Sail*, pp. 14–21.

[15]"Winging It," *The Economist*, September 18, 2004.

[16]Newton, *Encyclopedia of Air*, p 228.

[17]Roger Hamilton, "Can We Harness the Wind?" *National Geographic*, December 1975, p. 812.

[18]The power of the wind passing perpendicularly through a circular area is: $P = 1/2 \, v3 \, r2$, where P = the power of the wind measured in watts; v = the velocity of the wind measured in meters per second; r = the radius of the rotor measured in meters (from www.windpower.org).

[19]Vaitheeswaran, *Power to the People*, p. 130.

[20]Gipe, *Wind Energy Comes of Age*.

[21]Martin Mittelstaedt, "Who Has Seen the Wind Power," *Globe and Mail*, January 4, 2003.

[22]Alex Markels, "Prevailing Winds," *Mother Jones*, July-August 2002, p. 38.

[23]John Vidal, "Eye of the Storm," *The Guardian*, May 28, 2004.

[24]Katharine Seelye, "Windmills Sow Dissent for Environmentalists," *New York Times*, June 5, 2003.

[25]Vidal, *The Guardian* (see note 23, above).

[26]Bill McKibben, "It's Easy Being Green," *Mother Jones*, July-August 2002, p. 36.

[27]This summary of the Nantucket debacle, and these quotes, are from a *New York Times Magazine* piece by Elinor Burkett. Elinor Burkett, "A Mighty Wind," June 15, 2003.

[28]American Wind Energy Association Web site, February 2001. www.awea.org.

[29]Fiona Davies, "The Power of Water," *Corporate Knights* magazine, water issue, summer 2004.

[30]Vaitheeswaran, *Power to the People*, p. 247.

[31]Burkett, *New York Times Magazine* (see note 27, above).

EPILOGUE

Peter Bowyer interview, November 2004.

SELECTED
BIBLIOGRAPHY

The following books are cited in the text or the endnotes. Journal, magazine, and Web site references are sourced in the notes themselves.

Anon. *Relation of the Voyage to Siam Performed by Six Jesuits Sent by the French King to the Indies and China in the Year 1685*. London: printed by T. B. for J. Robinson and A. Churchill, 1688.

Bacon, Sir Francis. *Historia Ventorum* (Part III of his Instauratio Magna), online edition in English at www.sirbacon.org/naturalhistorywinds.htm

Bowyer, Peter, editor, *Where the Wind Blows*. Breakwater Books, St John's, 1995 & Environment Canada, Halifax.

Carrier, Jim. *The Ship and the Storm: Hurricane Mitch and the Loss of the Fantome*. New York: Harcourt, 2001.

Conrad, Joseph. *Typhoon and Other Stories*. London: Penguin Classics, 1990.

Cowley, Robert, editor. *What If? Eminent Historians Imagine What Might Have Been*. New York: Putnam, 2001.

DeBlieu, Jan. *Wind: How the Flow of Air Has Shaped Life, Myth, and the Land*. New York: Mariner Books, Houghton Mifflin, 1999.

Defoe, Daniel. *The Storm*. New York: Penguin Classics, 2005.

Diamond, Jared. *Collapse: How Societies Choose to Fail or Succeed*. New York: Viking Penguin, 2005.

Dorson, Richard M. *Buying the Wind: Regional Folklore in the United States*. Chicago: University of Chicago Press, 1964.

Dyson, John. *Spirit of Sail: On Board the World's Great Sailing Ships*. Toronto: Key Porter Books, 1987.

Fortey, Richard. *The Earth: An Intimate History*. London: HarperCollins, 2004.

Frazer, Sir James George. *The Golden Bough: A Study in Magic and Religion*. New York: Oxford University Press, 1994.

Gipe, Paul. *Wind Energy Comes of Age*. New York: John Wiley and Sons, 1995.

Heatter, Basil. *Eighty Days to Hong Kong: The Story of the Clipper Ships*. Toronto: Doubleday, 1969.

Herodotus. *Histories, Persian Wars*. Vol 2, Oxford: Oxford University Press, 1962.

Huler, Scott. *Defining the Wind: The Beaufort Scale, and How a 19th-Century Admiral Turned Science into Poetry*. New York: Crown, 2004.

Johnson, Donald S. *Phantom Islands of the Atlantic*. Fredericton, New Brunswick: 1994.

Junger, Sebastian. *The Perfect Storm*. New York: Norton, 1997.

Lamb, Hubert H. *Climate, History and the Modern World*. London: Methuen, 1995.

Larousse Encyclopedia of Mythology. London: Paul Hamlyn, 1969.

Larson, Erik. *Isaac's Storm: A Man, a Time, and the Deadliest Hurricane in History*. New York: Vintage, 2000.

Lively, Penelope. *Heat Wave*. London: HarperCollins, 1996.

Lomborg, Bjørn. *The Skeptical Environmentalist*. Cambridge: Cambridge University Press, 2004.

McKibben, Bill. *The End of Nature*. New York: Anchor Books, 1999.

Meadows, Donella, Jorgen Randers, and Dennis Meadows. *Limits to Growth: The 30-year Update*. White River Junction, VT: Chelsea Green, 2004.

Menzies, Gavin. *1421: The Year China Discovered America*. New York: William Morrow, 2001.

Newton, David E. *Encyclopedia of Air*. Westport, CT: Greenwood Press, 2003.

Pliny the Elder, *Natural History of the World*. Translated by Philemon Holland. London, 1601.

Proulx, E. Annie. *The Shipping News*. New York: Simon and Schuster, 1993.

Quarrington, Paul. *Galveston*. Toronto: Random House, 2004.

Reiss, Bob. *The Coming Storm: Extreme Weather and Our Terrifying Future*. New York: Hyperion, 2001.

Seth, Dunn. *Hydrogen Futures: Toward a Sustainable Energy System*. Washington, D.C.: Worldwatch Institute, 2001.

Sheets, Bob, and Jack Williams. *Hurricane Watch: Forecasting the Deadliest Storms on Earth*. New York: Vintage, 2001.

Smith, Sebastian. *Southern Winds*. London: Penguin, 2004.

Tylor, Sir Edward Burnett. *Primitive Culture: Researches into the Development of Mythology, Philosophy, Religion, Art, and Custom*. London: John Murray, 1871.

Vaitheeswaran, Vijay. *Power to the People*. New York: Farrar, Straus and Giroux, 2003.

Van der Post, Laurens. *A Story Like the Wind*. Toronto: Clarke Irwin, 1972.

Williams, Jack. *USA Today's Weather Book*. New York: Vintage, 1992.

Wright, Ronald. *A Short History of Progress*. Toronto: Anansi, 2004.

Young, Louise B. *Sowing the Wind: Reflections on the Earth's Atmosphere*. New York: Prentice Hall, 1990.

Zebrowski, Ernest Jr. *Perils of a Restless Planet: Scientific Perspectives on Natural Disasters*. Cambridge: University of Cambridge Press, 1997.

INDEX